T0332739

Machine Learning and Cognitive Science Applications in Cyber Security

Muhammad Salman Khan
University of Manitoba, Canada

A volume in the Advances in
Computational Intelligence and
Robotics (ACIR) Book Series

Published in the United States of America by
    IGI Global
    Information Science Reference (an imprint of IGI Global)
    701 E. Chocolate Avenue
    Hershey PA, USA 17033
    Tel: 717-533-8845
    Fax: 717-533-8661
    E-mail: cust@igi-global.com
    Web site: http://www.igi-global.com

Library of Congress Cataloging-in-Publication Data

Names: Khan, Muhammad Salman, 1979- editor.
Title: Machine learning and cognitive science applications in cyber security
    / Muhammad Salman Khan, editor.
Description: Hershey, Pa : Information Science Reference, [2019] | Includes
    bibliographical references.
Identifiers: LCCN 2018046997| ISBN 9781522581000 (h/c) | ISBN 9781522581017
    (eISBN)
Subjects: LCSH: Internet--Security measures. | Computer security--Data
    processing. | Machine learning. | Computer crimes--Prevention.
Classification: LCC TK5105.59 .M328 2019 | DDC 006.3/1--dc23 LC record available at https://
lccn.loc.gov/2018046997

This book is published in the IGI Global book series Advances in Computational Intelligence and
Robotics (ACIR) (ISSN: 2327-0411; eISSN: 2327-042X)

British Cataloguing in Publication Data
A Cataloguing in Publication record for this book is available from the British Library.

All work contributed to this book is new, previously-unpublished material.
The views expressed in this book are those of the authors, but not necessarily of the publisher.

For electronic access to this publication, please contact: eresources@igi-global.com.

Advances in Computational Intelligence and Robotics (ACIR) Book Series

ISSN:2327-0411
EISSN:2327-042X

Editor-in-Chief: Ivan Giannoccaro, University of Salento, Italy

MISSION

While intelligence is traditionally a term applied to humans and human cognition, technology has progressed in such a way to allow for the development of intelligent systems able to simulate many human traits. With this new era of simulated and artificial intelligence, much research is needed in order to continue to advance the field and also to evaluate the ethical and societal concerns of the existence of artificial life and machine learning.

The **Advances in Computational Intelligence and Robotics (ACIR) Book Series** encourages scholarly discourse on all topics pertaining to evolutionary computing, artificial life, computational intelligence, machine learning, and robotics. ACIR presents the latest research being conducted on diverse topics in intelligence technologies with the goal of advancing knowledge and applications in this rapidly evolving field.

COVERAGE

- Machine Learning
- Computer Vision
- Brain Simulation
- Computational Logic
- Synthetic Emotions
- Fuzzy Systems
- Robotics
- Artificial Life
- Natural Language Processing
- Heuristics

IGI Global is currently accepting manuscripts for publication within this series. To submit a proposal for a volume in this series, please contact our Acquisition Editors at Acquisitions@igi-global.com or visit: http://www.igi-global.com/publish/.

Titles in this Series

701 East Chocolate Avenue, Hershey, PA 17033, USA
Tel: 717-533-8845 x100 • Fax: 717-533-8661
E-Mail: cust@igi-global.com • www.igi-global.com

Table of Contents

Detailed Table of Contents

Chapter 1

Brian S. Coats, University of Maryland – Baltimore, USA
Subrata Acharya, Towson University, USA

Integrity, efficiency, and accessibility in healthcare aren't new issues, but it has been only in recent years that they have gained significant traction with the US government passing a number of laws to greatly enhance the exchange of medical information amidst all relevant stakeholders. While many plans have been created, guidelines formed, and national strategies forged, there are still significant gaps in how actual technology will be applied to achieve these goals. A holistic approach with adequate input and support from all vital partakers is key to appropriate problem modeling and accurate solution determination. To this effect, this research presents a cognitive science-based solution by addressing comprehensive compliance implementation as mandated by the Health Insurance Portability and Accountability Act, the certified Electronic Health Record standard, and the federal Meaningful Use program. Using the developed standardized frameworks, an all-inclusive technological solution is presented to provide accessibility, efficiency, and integrity of healthcare information security systems.

Chapter 2

Siu Cheung Ho, The Hong Kong Polytechnic University, Hong Kong
Kin Chun Wong, The Hong Kong Polytechnic University, Hong Kong
Yuen Kwan Yau, The Hong Kong Polytechnic University, Hong Kong
Chi Kwan Yip, The Hong Kong Polytechnic University, Hong Kong

Currently, Chinese commercial banks are facing extremely tremendous pressure, including financial disintermediation, interest rate marketization, and internet finance. Meanwhile, increasing financial consumption demand of customers further intensifies the competition among commercial banks. Hence, it is very important to store, process, manage, and analyze the data to extract knowledge from the customer to predict their investment direction in future. Customer retention and fraud detection are the main information for the bank to predict customer behavior. It may involve the privacy data and sensitive data of the customer. Data security and data protection for the machine learning prediction is necessary before data collection. The research is focused on two parts: the first part is data security of machine learning and second part is machine learning prediction. The result is to prove the data security for the machine learning is important. Using different machining learning analysis tool to enhance the performance and reliability of machine learning applications, the customer behavior prediction accuracy can be enhanced.

Machine learning is a field that is developed out of artificial intelligence (AI). Applying AI, we needed to manufacture better and keen machines. Be that as it may, aside from a couple of simple errands, for example, finding the briefest way between two points, it isn't to program more mind boggling and continually developing difficulties. There was an acknowledgment that the best way to have the capacity to accomplish this undertaking was to give machines a chance to gain from itself. This sounds like a youngster learning from itself. So, machine learning was produced as another capacity for computers. Also, machine learning is available in such huge numbers of sections of technology that we don't understand it while utilizing it. This chapter explores advanced-level security in network and real-time applications using machine learning.

The chapter introduces the perspectives on the use of avatar-based management techniques for designing new tools to improve blockchain as technology for cyber security issues. The purpose of this chapter was to develop an avatar-based closed model with strong empirical grounding that provides a uniform platform to address issues in different areas of digital economy and creating new tools to improve blockchain technology using the intelligent visualization techniques. The authors show the essence, dignity, current state, and development prospects of avatar-based management using blockchain technology for improving implementation of economic solutions in the digital economy of Russia.

This chapter is devoted to studying the opportunities of machine learning with avatar-based management techniques aimed at optimizing threat for cyber security professionals. The authors of the chapter developed a triangular scheme of machine learning, which included at each vertex one participant: a trainee, training, and an expert. To realize the goal set by the authors, an intelligent agent is included in the triangular scheme. The authors developed the innovation tools using intelligent visualization techniques for big data analytic with avatar-based management in sliding mode introduced by V. Mkrttchian in his books and chapters published by IGI Global in 2017-18. The developed algorithm, in contrast to the well-known, uses a three-loop feedback system that regulates the current state of the program depending on the user's actions, virtual state, and the status of implementation of available hardware resources. The algorithm of automatic situational selection of interactive software component configuration in virtual machine learning environment in intelligent-analytic platforms was developed.

### Chapter 6

The literature review of known sources forming the theoretical basis of calculations on Sleptsova networks and on the basis of authors' developments in machine learning with avatar-based management established the basis for the future solutions to hyper-computations to support cyber security applications. The chapter established that the petri net performed exponentially slower and is a special case of the Sleptsov network. The universal network of Sleptsov is a prototype of the Sleptsov network processor. The authors conclude that machine learning with avatar-based management at the platform of the Sleptsov net-processor is the future solution for cyber security applications in Russia.

### Chapter 7

Computers generate a large volume of logs recording various events of interest. These logs are a rich source of information and can be analyzed to extract various insights about the system. However, due to its overwhelmingly large volume, logs are often mismanaged and not utilized effectively. The goal of this chapter is to help researchers and industrial professionals make more informed decisions about their logging solutions. It first lays the foundation by describing log sources and format. Then it describes all the components involved in logging. The remainder of the chapter provides a survey of different log analysis techniques and their applications, consisting of conventional techniques using rules and event correlators that can detect known issues, plus more advanced techniques such as statistical, machine learning, and deep learning techniques that can also detect unknown issues. The chapter concludes describing the underlying concepts of the techniques, their application to log analysis, and their comparative effectiveness.

### Chapter 8

Over the years, passwords have been our safeguards by acting to prevent one's data from unauthorized access. With the advancement of technologies, the way we have been using passwords has changed and transformed into much secure yet more user friendly than they were ever been in the past. However, the vulnerabilities identified and observed in this traditional system has motivated industry and researchers to find some alternate where there is no threat like stealing, hacking, and cracking of password. This chapter discusses the major developed password-less authentication techniques in detail and also puts an effort to explain the in-depth details along with the working principle of each of the technique through a use-case diagram. It would be of great benefit and contribution to the callow trying to explore research opportunities in this area.

Mihoubi Miloud, Djillali Liabes University, Algeria
Rahmoun Abdellatif, École Supérieure en Informatique, Algeria
Pascal Lorenz, University of Haute Alsace, France

WSNs have recently been extensively investigated due to their numerous applications where processes have to be spread over a large area. One of the important challenges in WSNs is secure node localization. Its main objective is to protect the circulated information in WSN for any attack with low energy. For this reason, recent approaches relying on swarm intelligence techniques are called and the node localization is seen as an optimization problem in a multi-dimensional space. In this chapter, the authors present an improvement to the original bat algorithm for information protecting during the localization task. Hence, the proposed approach computes iteratively the position of the nodes and studied the scalability of the algorithm on a large WSN with hundreds of sensors that shows pretty good performance. Moreover, the parameters are simulated in different scenarios of simulation. In addition, a comparative study is conducted to give more performance to the proposed algorithm.

Rajakumar Arul, Amrita Vishwa Vidyapeetham, India
Rajalakshmi Shenbaga Moorthy, St. Joseph's Institute of Technology, India
Ali Kashif Bashir, Manchester Metropolitan University, UK

Technology evolution in the network security space has been through many dramatic changes recently. Enhancements in the field of telecommunication systems invite fruitful security solutions to address various threats that arise due to the exponential growth in the number of users. It's crucial for upgrading the entire infrastructure to safeguard the system from specific threats. So, there is a huge demand for the learning mechanism to realize the behavior of attacks. Recent upcoming technologies like machine learning and deep learning can support in the process of learning the behavior of all types of attacks irrespective of their deployment criteria. In this chapter, the analysis of various machine learning algorithms with respect to a few scenarios that can be adopted for the benefits of improving the security standard of the network. This chapter briefly discusses various know attacks and their classification and how machine learning algorithms can be involved to overcome the popular attacks. Also, various intrusion detection and prevention schemes were discussed in detail.

Preface

With the ever-increasing complexity of cyber threat landscape, existing cyber defense methods and technologies are being faced with the challenges of not only fading effectiveness of state of the art threat detection mechanisms, but efficiency has also become a major challenge in the context of growing false alerts which is further compounded by information overload for cyber experts. Moreover, knowing these weaknesses of existing cyber technologies, including but not limited to IDS, IPS, complex password management systems, SIEMs and firewalls, cyber adversaries have increased intelligence of their attack tactics by opening multiple cyber kill chains which are aimed at diverting the attention of cyber defenders and thus provide a way of breaching the system with advanced malware or hacking tools and subsequently exploiting information overload and distraction.

Machine learning tools have shown a promising way to not only automate the threat detection and investigation tasks (such as playbook scenarios for threat detection, orchestration and hunting), but also provided a reasonable mechanism to reduce the cognitive load of human cyber experts. However, with the increasing complexity of cyber world such as complex connectivity patterns, various abstraction layers of data flows and myriad of configurations, it is becoming more difficult to manage cyber infrastructure. Further, with the increasing complexity of attack vectors and malware techniques to evade existing cyber defense mechanisms, it is required to build new machine learning tools with sufficient intelligence to detect and respond effectively and efficiently. Cognitive machine intelligence is a new topic that covers not only machine learning and deep learning tools but also provide abstract mechanisms to improve the threat detection capabilities and performance of these tools. It is paramount to understand that cyber issues are very unique and differentiate from other application domains such as image processing, health care and financial world. The primary differentiating factor are high speed data, lack of structured patterns and volatility factors which account to change in machine intelligence approaches from statistical and application perspective. For example, it is not possible to only look for outliers using machine learning and anomaly detection systems, since internet traffic and application data have evidently reported heavy tailed distribution that warrants the research and development of machine learning in new directions.

Machine Learning and Cognitive Science Applications in Cyber Security will introduce the readers with various aspects of machine learning and cognitive machine intelligence to address variety of cyber security application challenges such as computation optimization, cryptography, analysis of logs and factors required to analyze health care data. This book introduces the concepts, models, algorithms and solutions pertaining to various aspects of cyber security and enrich the readers with the required knowledge and skills necessary to excel in the application domain of cognitive machine intelligence in the world of cyber security. Moreover, this book also provides a taste of the current state of research which is aimed in the application of machine learning tools in cyber security and what could be the future directions to combat the ever growing complexity of cyber threats.

Muhammad Salman Khan
University of Manitoba, Canada

Acknowledgment

The editors are grateful for the assistance of the following fellows, who reviewed the manuscripts and provided their valuable feedback. The editors are also thankful to IGI Global staff particularly Josephine Dadeboe, Alyssa Reimenschneider, Jan Travers, and Marianne Caesar.

Rajakumar Arul, India

Rahul Singh Chowhan, India

Dr. Vardan Mkrttchian, Australia

Dr. Melody Moh, USA

Rajalakshmi Shenbaga Moorthy, India

Dr. Mamata Rath, USA

Mohammad Nurul Afsar Shaon, Canada

Chapter 1
Healthcare Information Security in the Cyber World

Brian S. Coats
University of Maryland – Baltimore, USA

Subrata Acharya
Towson University, USA

ABSTRACT

Integrity, efficiency, and accessibility in healthcare aren't new issues, but it has been only in recent years that they have gained significant traction with the US government passing a number of laws to greatly enhance the exchange of medical information amidst all relevant stakeholders. While many plans have been created, guidelines formed, and national strategies forged, there are still significant gaps in how actual technology will be applied to achieve these goals. A holistic approach with adequate input and support from all vital partakers is key to appropriate problem modeling and accurate solution determination. To this effect, this research presents a cognitive science-based solution by addressing comprehensive compliance implementation as mandated by the Health Insurance Portability and Accountability Act, the certified Electronic Health Record standard, and the federal Meaningful Use program. Using the developed standardized frameworks, an all-inclusive technological solution is presented to provide accessibility, efficiency, and integrity of healthcare information security systems.

DOI: 10.4018/978-1-5225-8100-0.ch001

INTRODUCTION

Healthcare providers and payers have been attempting to achieve HIPAA compliance for nearly a decade. In 1998, shortly after HIPAA's signing, the research firm Gartner Group estimated the implementation of HIPAA would collectively cost healthcare providers $5 billion and health plans $14 billion. By 2005, HHS was estimating that the costs could be at least 3 times the original amount for providers and as much as 10 times the original amount for health plans (HIPAA Security Rule, 2008). In 2009, HIMSS sponsored research suggested that the actual implementation costs for providers would be closer to $40 billion (Title 45-Public Welfare, 1996). This trend indicates a considerable cost increase that in some cases could prove crippling, especially for smaller entities. The costs of these implementations have deviated even more than their timelines and creating financial burdens drastically higher than originally anticipated. Surmounting costs aside, the original schedule set by the Privacy and Security Rules required compliance by 2003 and 2005 respectively (EHR Adoption Trends, 2004). Clearly these compliance goals have not been met by most healthcare organizations around the country. While the road to HIPAA compliance is proving elusive and costly, organizations clearly understand the importance and necessity of completing the undertaking. HIPAA will ultimately ensure better privacy and security of ePHI data. Organizations have both ethical and financial motivations to provide their customers the guarantees that HIPAA requires and are spending massive amounts of time and money on their implementations. It is critical for these organizations to have clear and comprehensive guidelines to follow for maximum efficiency in their efforts.

There are a variety of reasons why HIPAA implementations have proved more expensive and taken considerably longer than originally anticipated by federal regulators and healthcare organizations alike. The biggest hurdle to overcome is simply the creation of an assessment, testing, and implementation plan. While many government agencies, private foundations, and industry consortiums have established high level guidelines and recommendations of how to address each of the HIPAA Rules, there is no nationally mandated implementation plan or standardized framework for organizations to follow. Each entity is responsible for reviewing the guidelines and determining the appropriate solution. The published recommendations are at a very abstract level and require much interpretation to formulate an actual implementation strategy. With a lack of clear direction, many entities have difficulty determining the best path for them to follow to satisfy each requirement. Furthermore, without an apparent plan or timeline, it becomes extremely difficult for organizations to generate realistic cost estimates for their compliance efforts and likewise secure the necessary budgetary commitments. This point has been demonstrated consistently since the first HIPAA implementations began. National cost estimates of HIPAA efforts are approaching a factor of ten higher than what regulators estimated when the law was first enacted (Coats, Acharya, Saluja, and Fuller, 2012).

One of the major steps towards fully meeting the HIPAA regulations is the implementation of an EHR system. With over 90% of healthcare providers in some stage of an EHR solution, HIMSS indicates that as of December 2011, only 66 hospitals, just over 1% nationally, have actually achieved Stage 7 – the final EHR adoption stage (Blumenthal and Tavenner, 2010). Furthermore, even with the federal government offering anywhere from $100,000 to over $2 million per provider, per year just to demonstrate the 'meaningful use' of a partial EHR implementation, only about 41% of providers have cashed in. Over $5.5 billion has already been paid to healthcare providers participating in the Meaningful Use program, but almost another potential $8 billion is being left unclaimed. Clearly providers are being given the proper motivation to implement EHR systems but are finding themselves ill-equipped to take the necessary steps to accomplish the task.

EHR systems will afford significant cost savings to healthcare providers by streamlining and standardizing their exchange and storage of ePHI. These systems will also enable better access to patient data by all parties - providers, insurers, and patients themselves. But with this improved access, healthcare providers are presented with the challenge of ensuring both privacy and security are preserved. Additionally, providers have the daunting task of making the process of patients gaining electronic access to their data simple and straightforward. The healthcare industry, like all industries, entered the digital age with each provider creating its own silos of data stores and corresponding security frameworks to access that data. As such, they are finding themselves poorly positioned to enable the distributed access to their data that EHR systems facilitate. The regulations and programs, including HIPAA and Meaningful Use, that are driving EHR adoption, provide almost zero guidance on how to address these enormous usability issue. The federal program NSTIC is singularly tasked with creating an "Identity Ecosystem" of interoperable technology standards and policies to provide increased security and privacy, but most importantly ease of use for individuals (EHR Incentive Programs, 2012). The Department of Health and Human Services, the agency responsible for HIPAA and Meaningful Use, is intimately involved in the development of NSTIC. This strategy will force the healthcare entirely restructure their approaches to identity access and management from centralized to distributed models.

The recognition of the need for a standardized framework for healthcare information security is widespread throughout the healthcare industry and federal government, all the way to the White House. Whether it be HIPAA compliance, EHR systems and Meaningful Use, or distributed electronic patient access, every healthcare entity is approaching these issues independently. This approach is continuing to prove both costly and timely, and ultimately the general public feels the impact. The fundamental objectives of all these regulations and programs provide for valuable improvements to the overall health care in the United States. Unfortunately, these benefits can only be realized when the programs are completed and at present that necessary steps are proving extremely challenging for healthcare organizations.

To this effect, the goal of this research is to bridge the gap from regulation to practice in a number of key technological areas of healthcare information security. Using standardized frameworks, this research proposes how accessibility, efficiency, and integrity in healthcare information security can be achieved. This research will converge on these issues by addressing HIPAA compliance, EHR Adoption and the federal Meaningful Use program, and pervasive electronic access for patients; all from the healthcare provider's perspective as shown in Figure 1.

The salient contributions of this research are:

- *The creation of the Healthcare Information Security Framework (HISF) to offer direction for organizations to plan and execute their overall HIPAA compliance projects including attestation,*
- *The creation of the Healthcare Information Security Plan (HISP) to provide comprehensive implementation level guidance for satisfying HIPAA regulations,*
- *The creation of the Healthcare Information Security Testing Directive (HISTD) and a collection of open source security testing software for organizations to assess and mitigate their systems.*
- *The creation of the Healthcare Federated Identity Framework (HFIF) that will position healthcare providers to enable distributed electronic access to patient data.*

Figure 1. Standardized model

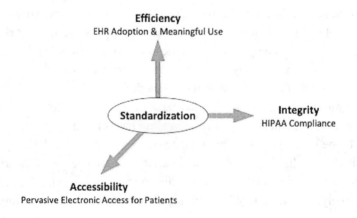

Background and Related Work

Information security has perpetually been a hot topic for all industries. The specific subject of healthcare information technology (HIT) and healthcare information security (HIS) has sparked a vast amount of research over the last few decades and is reflected in a wide array of peer-reviewed scholarly papers and journal articles. Furthermore, much attention has been given to the difficulties faced in HIT and HIS related to Health Insurance Portability and Accountability Act (HIPAA) implementation and assessment, EHR adoption, and patient accessibility. After a thorough examination of a substantial amount of the related literature, clear shortcomings became evident in the technological solutions as many researchers lamented some common problems and searched for answers.

Healthcare providers and payers have been attempting to achieve Health Insurance Portability and Accountability Act (HIPAA) compliance for nearly a decade. In 1998, shortly after HIPAA's signing, the research firm Gartner Group estimated the implementation of HIPAA would collectively cost healthcare providers $5 billion and health plans $14 billion nationally. As early as 2003 when the final regulations for both HIPAA rules had been released, healthcare legal expert George Annas (Annas, 2003) was already predicting HIPAA implementations to be "costly, inconsistent, and frustrating". Annas went on to state that "HIPAA consultants" were quickly becoming necessary for hospitals, health plans, and physician practices in order to understand how to comply with "long, complex" unclear regulations. By 2009, researchers still echoed that sentiment by offering that providers have pressure to hire external consultants as "there is a high degree of uncertainty associated with the interpretation of regulations and organizations lack adequate in-house resources" (Appari, Anthony, and Johnson, 2009). The Department of Health & Human Services (HHS) was already estimating that the costs could be at least 3 times the original amount for providers and as much as 10 times the original amount for health plans (HIPAA Return on Investment, 2005). By 2009, HIMSS sponsored research suggested that the actual nationwide implementation costs for providers would be closer to $40 billion. Most recently, the Department of Health and Human Services had re-estimated this figure to have grown to a national average of $114 million to $225 million in the first year and a recurring annual cost of $14.5 million, per healthcare provider (HIPAA Final Rule, 2013). This trend indicates a considerable cost increase that in some cases could prove crippling, especially for smaller entities. The costs of these implementations have deviated even more than their timelines and are creating financial burdens drastically higher than originally anticipated. Many providers and researchers argue the HHS is still significantly underestimating the actual compliance costs (Hirsch, 2013).

Surmounting costs aside, the original schedule set by the Privacy and Security Rules required compliance by 2003 and 2005 respectively (Regulations and Guide, 2014). Unfortunately, these compliance goals were not met by most healthcare organizations around the country. In 2008, the Centers for Medicare and Medicaid Services performed a review of HIPAA covered entities (CEs) and their compliance only with the Security Rule. This review demonstrated that CEs continued to struggle with meeting all aspects of the regulations, specifically in the areas of risk assessment, currency of policies and procedures, security training, workforce clearance, workstation security, and encryption (HIPAA Compliance Review Analysis and Summary of Results, 2008). Even in 2013, Solove (2013) discusses the still present gap in HIPAA compliance by all CEs and offers that "in addition to the dynamism of HIPAA, compliance is not something that is ever completely solved". He continues that it is not a one-time implementation but rather a daily challenge to maintain. As such, healthcare entities are faced not only with just becoming HIPAA compliant at single point in time, but they must achieve and maintain compliance perpetually. Therefore, it is critical for these organizations to have clear and comprehensive guidelines to follow for maximum efficiency in their efforts.

Fichman, Kohli and Krishnan (2011) note that because ePHI is personal by nature this compounds public fears and concerns related to data breaches. Healthcare providers must work hard to gain the trust of their patients and work even harder to maintain that trust. Data breaches have severe consequences for providers ranging from fines, embarrassment, reputational damage, and remediation costs (Kwon and Johnson, 2013). HHS, as part of the Health Information Technology for Economic and Clinical Health (HITECH) Act, has implemented a new data breach notification process that requires healthcare entities to publicly post breach announcements for cases involving 500 or more individuals (Breach Notification Rule, 2009). Similarly, HITECH also increased the severity of the fines up to $1.5 million for HIPAA violations related to both inadvertent and willful disclosure of patient data. While the penalties are dramatically increasing and organizations are investing in security protection and assessment tools, the reality is there is still a significant gap between the regulations and practice. In the first two years since the HHS installed the Breach Notification Rule, over 10 million patients' data were inappropriately disclosed (Enforcements Results per Year, 2010). A number of issues have been identified as the reason for healthcare organization's limited success with implementing security practices that are effective and compliant with the HIPAA directives. These issues include superficial implementations that don't align technology and business practices (Kayworth and Whitten, 2010), the tendency to be reactive instead of proactive due lack of active, established security programs and security measures being implemented piece meal instead of a comprehensive, complimentary approach (Xia and Johnson, 2010). Compliance as "a snapshot of security about whether an organization exhibits controls". They offer that organizations are more driven by compliance than true data security. Johnson, Goetz and Pfleeger (2009) cautioned that organizations that employ security assessment models with a "check-the-box" mentality do not have true assurance their security measures are effective; it is only through comprehensive testing and auditing that the measures are vetted. Aral and Weill (2007) make the apt distinction that actual security is defined by how well the security controls used for compliance are deployed and function.

As much of the published literature confirms, the core challenge that healthcare providers face with meeting Health Insurance Portability and Accountability Act (HIPAA) compliance, while also ensuring effective security, is simply the creation of a plan to assess and test their environments. Further, once the assessments and tests are complete, the organizations also need a remediation plan in the form of an implementation guide to react to any issues discovered. In an effort to provide organizations a standardized approach for addressing the HIPAA regulations, the National Institute for Standards and Technology (NIST) produced special publication 800-66 that focused on the implementation of the HIPAA Security Rule. This guide gets closer to the concept of mapping regulation to implementation but still does not provide specific actionable recommendations. While many government agencies, private foundations, and industry consortiums have established high level guidelines and recommendations of how to address each of the HIPAA rules, there is no nationally mandated implementation plan or standardized framework for organizations to follow. Each entity is responsible for reviewing the guidelines and determining the appropriate solution.

The idea of having actionable plans based off these various publications as well as other industry best practices is not a novel concept in of itself (Acharya, Coats, Saluja, and Fuller, 2014). The Health Information Trust (HITRUST) Alliance has created their *Common Security Framework (CSF)* to serve as a holistic solution to this significant need. HITRUST presents their CSF as a "comprehensive and flexible framework that remains sufficiently prescriptive in how control requirements can be scaled and tailored for healthcare organizations of varying types and sizes". Furthermore, the CSF includes federal regulations and standards such as HIPAA, Payment Card Industry Data Security Standards (PCI DSS), and Control Objectives for Information and Related Technology (COBIT) as well as recommendations from NIST, the Federal Trade Commission (FTC), and the International Organization for Standardization (ISO). The scope of the CSF in fact exactly matches the need of a prescriptive, standardized solution for healthcare organizations to follow. As such, it is not surprising that the CSF is the most widely adopted security framework by the healthcare industry in the United States. Unfortunately, the CSF like the consulting firms, comes with a substantial price tag for an annual subscription to access their framework content and information and have very limited (ranges from 10-20 annually depending on subscription tier) 'tickets' for working with a knowledgeable professional about how to implement the CSF.

Outside of healthcare, the concept of establishing standardized frameworks is very common. NIST has established the Risk Management Framework (RMF) to provide a systematic approach for managing organizational risk across all industries and sectors. The framework can be applied to either new or existing information systems to evaluate risk as well select, implement, assess, and monitor mitigating controls to risk. Similarly, the Financial Services Roundtable, a collaborative body made up of the leadership of the nation's largest financial institutions, saw the need to create a standardized approach for information security within the financial industry. As a result, the verbose Banking Industry Technology Secretariat (BITS) Security Program was created that shared information security best practices and successful strategies.

Up to this point a comprehensive solution, like the CSF, RMF, or BITS Security Program, has not been presented in an open academic format for the healthcare industry such that organizations can perform both the abstract style assessment using questionnaires and surveys as well as conduct the active penetration testing themselves. What is also missing from the current commercial offerings is the ability to see specifically the derivation of the all the assessment mechanisms so that they can be updated and adapted if and when regulations are added or changed. This mapping information, tying regulation to practice and assessment, is proprietary to the commercial offerings as it effectively constitutes the entire value of their engagements aside from the man-hours to perform the assessment. Therefore, as it stands today, 2 basic options have developed, either 'pay to play' by contracting with one of the private security assessment firms that specialize in HIPAA compliance or establish a subscription to HITRUST's CSF, or alternately use the NIST guideline and muddle through alone. With many organizations' considerable budget constraints, the latter option of proceeding independently using the existing guidance tends to become the common option. Additionally, without an apparent plan or timeline to follow, it becomes extremely difficult for organizations to generate realistic cost estimates for their compliance efforts and likewise secure the necessary budgetary commitments. This results in enormous wastes of capital, time, and energy for the healthcare provider. This point has been demonstrated consistently since the first HIPAA implementations began. Consequently, national cost estimates of HIPAA efforts have well eclipsed a factor of ten higher than what regulators estimated when the law was first enacted.

Only further complicating the HIPAA compliance landscape, the final rules of the Health Information Technology for Economic and Clinical Health (HITECH) Act of 2009 introduced significant changes to the prior HIPAA regulations. While designed to encourage the development of health information exchanges these changes are still requiring additional attention and therefore cost and effort to be extended to HIPAA compliance. As part of these changes, the rules expanded the types of entities that are covered by HIPAA. Previously, HIPAA only dealt with healthcare providers, health plans, and healthcare clearinghouses. The final rules released in January 2013 now defines covered entities as any vendor that creates, transmits, receives, or maintains ePHI. Furthermore, these additional entities can now be held civilly and criminally liable. Aside from new entities being covered, even those entities that had achieved or were close to achieving HIPAA compliance coming into 2013 are now having to evaluate and accommodate the considerable additions and changes to the regulations. The result of these recent changes has even more entities scrambling to become HIPAA compliant and effectively taking the cumulative percentage of all covered entities' compliance further away from the hundred percent target.

When examining the potential solutions for providing electronic patient access to EHRs there are numerous existing models to consider although this research submits that a viable solution has yet to be developed that is scalable, cost-effective, and easily available to virtually everyone. This electronic identity situation has many healthcare providers finding themselves poorly positioned to enable the types of distributed access that EHR systems are supposed to facilitate. The regulations and programs that are driving EHR adoption, including HIPAA and Meaningful Use, provide virtually no direction on how to tackle these enormous usability and efficiency challenges. The federal program National Strategy for Trusted Identities in Cyberspace (NSTIC) is aggressively working to establish interoperable technology standards and policies for sharing identity information potentially anywhere in the public or private sectors. The Department of Health and Human Services, the agency responsible for HIPAA and Meaningful Use, is intimately involved in the development of NSTIC. This strategy will compound the need for healthcare entities to entirely restructure their approaches for identity and access management from centralized to distributed models.

Framework and Methodology

Organizations have both ethical and financial motivations to provide their customers the guarantees and benefits that the Health Insurance Portability and Accountability Act (HIPAA) and EHR systems afford. As a result, healthcare providers are spending massive amounts of time and money on their implementations. When considering the 3 basic goals of this research - integrity, efficiency, and accessibility in healthcare - it became clear that any technology solution would require a delicate balance of these 3 areas in order to be viable for practical application. As such, integrity and accessibility quickly became the 2 pillars and motivations of the solutions, while efficiency became the measure of success. Integrity (or compliance) was the first of the 2 foundational elements tackled. In a very basic sense, the design approach was to determine how an organization could measure and achieve compliance (ensure integrity) in an efficient manner. Once integrity had been addressed, the research's attention shifted to how to make healthcare access more easily attainable and efficient. These efforts resulted in the creation of 2 unique frameworks that aim to bring together integrity, accessibility, and efficiency for a healthcare provider's organization. The process is included as a step-by-step manner in the Figure 2.

Healthcare Information Security Compliance Framework

A federal grant from the Department of Health & Human Services (HHS), that begun in early 2011, connected Towson University and a large federally-funded regional trauma center and national healthcare provider located in central Pennsylvania national healthcare provider (specific identity of the hospital has been suppressed due to non-disclosure agreement of grant). Part of the deliverables of this grant was to assess the Pennsylvania Hospital's HIPAA compliance and to respond to any shortcomings. As a result of this original need, the Healthcare Information Security Compliance Framework (HISCF) concept was first developed with the very specific goal of creating a HIPAA compliance assessment plan for a hospital. During the early discovery and research stage of the grant, it became very apparent that there was a clear lack of implementation level guidance on how to achieve HIPAA compliance and furthermore how to assess it. It became equally evident that the research and work associated with bringing this large national hospital into compliance could be leveraged to create and propose a standardized, reusable model that other organizations could potentially benefit from.

Framework Creation Process

With the goal of creating a standardized method for assessing an organization's HIPAA compliance and addressing any findings, it became apparent there were key

Figure 2. Conceptual view of proposed Framework

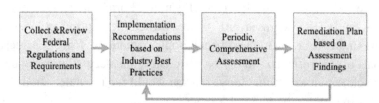

steps to accomplish this task. The first step was to create a comprehensive set of all the HIPAA regulations, consisting of the Security and Privacy Rules, as well as any other requirements laid out by HHS related to HIPAA, including the revisions to HIPAA spelled out in the Health Information Technology for Economic and Clinical Health (HITECH) Act. Once all the requirements had been defined, the next step was analyzing their technical implications and what implementation decisions would have relevance to compliance. Following this general analysis, research was done on what guidance NIST, HHS, and other federal agencies had provided to date, and what guidance private organizations like the Healthcare Information Management Systems Society (HIMSS) had produced both from a regulation and implementation perspective (Acharya, Coats, Saluja, and Fuller, 2013). At this point, all the regulations and requirements had been documented, their technical implications identified, and the existing guidance reviewed. It was during this step that the gap of actual implementation guidance was continually observed.

The next steps were to perform an examination of how HIPAA and other types of security and privacy assessments were being accomplished at other healthcare organization as well as and non-healthcare entities. This research formed the basis for the creation of the HISCF. The HISCF at its very core is an internal information security audit using the HIPAA regulations as the effective measurement of success or failure.

Figure 3. Healthcare information security compliance framework

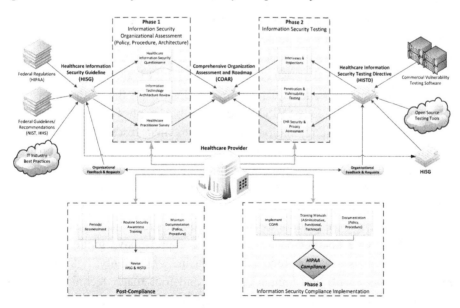

Starting with the conceptual basis shown in Figure 2, each step of the process was expanded into phases of actual tasks. The result was the formation of the framework shown in Figure 3. The proposed compliance framework consists of three primary phases enabling complete HIPAA compliance at its conclusion. The framework is designed to take an organization from the initial recognition of the need for compliance all the way through to implementation of any necessary changes to their environment. Further, the framework provides a post-compliance phase to ensure the healthcare provider maintains their compliance perpetually. While the phases and associated tasks are performed sequentially, there are feedback loops at almost every stage to reflect findings and feedback of successive steps to the preceding steps to ensure the assessment guides and instruments are organizationally relevant. NIST acknowledges a well-documented and repeatable compliance framework will greatly speed up the assessment and testing process, yield more consistent results, present less risk to the normal business operations of the organization, and minimize the resources needed to perform the testing. This research offers a comprehensive solution to organizational assessment and information security testing by providing step-by-step instructions for how to plan and perform information security compliance assessment and testing, how to analyze the results of the tests, and ultimately how to correct and mitigate any findings.

Healthcare Information Security Guide (HISG)

Many of the various activities laid out in the framework rely on the creation of the Healthcare Information Security Guide (HISG). The HISG is an invention of this research to serve as the cumulative, comprehensive reference manual for healthcare providers to use both for implementation assistance as well as later for assessment. In conjunction with the research performed while creating the HISCF, the basic content of the HISG was likewise compiled. The HISG is the culmination of the actual HIPAA regulations, federal recommendations from the National Institute for Standards and Technology (NIST) and the Department of Health and Human Services (HHS). Once the regulations were documented, research was performed to determine actual implementation suggestions that meet those regulations. The specific implementation recommendations incorporate standards and best practice guides provided by NIST, the National Security Agency (NSA), the Department of Homeland Security, and a myriad of public guides and private industry whitepapers.

Once the information, guidelines, and requirements from all these sources was compiled, they were distilled into a concise, comprehensive guide that covers four key policy areas - disaster recovery and business continuity; risk mitigation; operations management; and logical access - and four major technical areas of information technology - network; database; applications; and infrastructure. The

Figure 4. Regulations and requirements for information security compliance

Federal Government Agencies
National Archives and Records Administration
Federal Register, Vol. 78, No. 17, Part II – 45 CFR Parts 160 and 164
Department of Health and Human Services
42 CFR parts 412, 412, and 495: Medicare and Medicaid Programs; Electronic Health Record Incentive Program Stage 2; Health Information Technology: Standards, Implementation Specifications, and Certification Criteria for Electronic Health Record Technology, 2014 Edition; Revisions to the Permanent Certification Program for Health Information Technology; Final Rules
45 CFR parts 160 and 164: Modifications to the HIPAA Privacy, Security, Enforcement and Breach Notification Rules Under the Health Information Technology for Economic and Clinical Health Act and the Genetic Information Nondiscrimination Act: Other Modifications to the HIPAA Rules: Final Rule
Center for Medicare and Medicaid Services
"Regulations and Guidance," available at https://www.cms.gov/home/regsguidance.asp
"HIPAA Security Series – Security Standards: Technical Safeguards," available at http://www.hhs.gov/ocr/privacy/hipaa/administrative/securityrule/techsafeguards.pdf
"CMS System Security and e-Authentication Assurance Levels by Information Type," available at http://www.cms.gov/Research-Statistics-Data-and-Systems/CMS-Information-Technology/InformationSecurity/Downloads/System-Security-Levels-by-Information-Type.pdf
"CMS EHR Meaningful Use Overview," available at https://www.cms.gov/Regulations-and-Guidance/Legislation/EHRIncentivePrograms/Meaningful_Use.html
"Logical Access Controls and Segregation of Duties," available at http://www.cms.gov/Research-Statistics-Data-and-Systems/CMS-Information-Technology/InformationSecurity/downloads/WP02-Logical_Access.pdf
Office of Management and Budget
"M-04-04: E-Authentication Guidance for Federal Agencies," available at http://www.whitehouse.gov/sites/default/files/omb/memoranda/fy04/m04-04.pdf
National Institute of Standards and Technology
An Introductory Resource Guide for Implementing the Health Insurance Portability and Accountability Act (HIPAA) Security Rule (SP800-66 rev1)
"Risk Management Framework (RMF)" available at http://csrc.nist.gov/groups/SMA/fisma/framework.html

HISG then serves as the emblematic ruler that the healthcare organization is evaluated against and appropriate recommendations are derived from for the organization as a remediation plan for any shortcomings. Figure 4 and Figure 5 depict the key sources of regulations and requirements, as well as implementation recommendations and practices respectively.

Phase 1: Information Security Organizational Assessment

The goal of Phase 1 is to carry out a high-level assessment involving a thorough review of all policies, procedures, practices, and architectural designs. This stage

Figure 5. Implementation recommendations and practices

Federal Government Agencies
Department of Homeland Security - Federal Network Security Branch
"Continuous Asset Evaluation, Situational Awareness, and Risk Scoring Reference Architecture Report (CAESARS)," available at http://www.dhs.gov/xlibrary/assets/fns-caesars.pdf
National Institute of Standards and Technology
Guide for Conducting Risk Assessments (SP800-30 rev1)
Risk Management Guide for Information Technology Systems (SP800-37 rev1)
Managing Information Security Risk: Organization, Mission, and Information System View (SP800-39)
Creating a Patch and Vulnerability Management Program (SP800-40 ver2)
Guidelines on Firewalls and Firewall Policy (SP800-41 rev1)
Guidelines on Securing Public Web Servers (SP800-44 ver2)
Guide to Securing Legacy IEEE 802.11 Wireless Networks (SP800-48 rev1)
Guide for Assessing the Security Controls in Federal Information Systems and Organizations (SP800-53A rev1)
Recommendation for Key Management, Part 1: General (Draft SP800-57 part1 rev3)
Electronic Authentication Guide (SP800-63 rev1)
Guide to Intrusion Detection and Prevention Systems (IDPS) (SP800-94)
Technical Guide to Information Security Testing and Assessment (SP800-115)
Guidelines for Securing Wireless Local Area Networks (WLANs) (SP800-153)
National Security Agency – Central Security Service
"Security Configuration Guides" available at http://www.nsa.gov/ia/mitigation_guidance/security_configuration_guides/index.shtml
Office of Government-wide Policy
"Federal Identity, Credential, and Access Management," available at http://www.idmanagement.gov/pages.cfm/page/ICAM
Private Organizations
Healthcare Information Management Systems Society (HIMSS)
"Guidelines for Establishing Information Security Policies at Organizations with Computer-based Patient Record Systems," available at http://www.himss.org/content/files/CPRIToolkit/version6/v7/D38_CPRI_Guidelines-Information_Security_Policies.pdf
"HIMSS Application Security Questionnaire (HIMSS ASQ)," available at http://www.himss.org/content/files/ApplicationSecurityv2.3.pdf
Medical Universities
Johns Hopkins University
"Information Technology Policies," available at http://www.it.johnshopkins.edu/policies/itpolicies.html
University of California
"Guidelines for HIPAA Security Rule Compliance University of California," available at http://www.universityofcalifornia.edu/hipaa/docs/security_guidelines.pdf
State Governments
State of California. Office of Information Security
"California Information Security Risk Assessment Checklist (CA ISRAC)," available at http://www.cio.ca.gov/OIS/Government/documents/docs/RA_Checklist.doc
State of Maryland. Department of Information Technology
"Information Security Policy," available at http://doit.maryland.gov/support/Documents/security_guidelines/DoITSecurityPolicyv3.pdf
State of North Carolina. Statewide HIPAA Assessment Team
"North Carolina HIPAA Impact Determination Assessment (NC HIDA)," available at http://hipaa.dhhs.state.nc.us/hipaa2002/amicovered/doc/ImpactDeterminationQuestionnaire-Step2-2.doc

is broken into three parts - the Healthcare Information Security Questionnaire, the Information Technology Architecture Review, and the Healthcare Practitioner Survey. These three instruments are designed to measure information security compliance from both technical and functional perspectives. The grant's project director provided quality checks on the instruments to ensure their appropriateness and completeness for the areas the instruments were designed to assess - no external quality evaluation was performed.

Healthcare Information Security Questionnaire (HISQ)

Computing environments by their nature have intrinsic risks that require some form of mitigating action to minimize the potential for harm. These vulnerabilities are essentially any attribute or characteristic of the environment that can be exploited to violate established security policies or cause a deleterious effect. Organizations therefore should have vulnerability assessment plans that are executed routinely to detect, identify, measure, and understand the risks present in their information technology environments. The Healthcare Information Security Questionnaire (HISQ) is designed to comprehensively assess the organization's information security policies, procedures, and practices. The HISQ represents the bulk of the Phase 1 assessment as it evaluates the organization's compliance with the baseline requirements of the HISG. The questionnaire itself was designed by creating sets of dichotomous and semantic differential questions to determine how the organization's policies, procedures, and practices compared to those laid out in the HISG. The assessment is divided into the same 4 key policy subjects as well as 4 overarching technical areas described in the HISG.

There are four key policy areas that the HISQ examines in specific detail: Disaster Recovery and Business Continuity; Risk Management; Operations Management; and Logical Access. These aspects of information technology cut across an organization's strategic and operational practices. Both HIPAA and Meaningful Use clearly lay out numerous requirements in these critical areas. The policy sections of the HISQ are presented in the form of a questionnaire that in most cases asks straightforward, single choice answers. This area is typically completed by the healthcare provider's IT leadership or their representative as it covers the overall organization's IT policies and established procedures.

The technical assessment is likewise divided into the areas of Network, Application, Database, and Infrastructure. In contrast to the policy review, the technical sections are best completed by IT engineers or someone intimately familiar with the technical configuration of the organization IT environment. The technical sections allow for much more free form answers to accommodate and capture environment-specific details. Many of the questions posed in the technical section are directed at specific implementation choices and details compared to the more general inquiries of the policy and procedure sections. The results assist in providing a comprehensive evaluation of the entire technical architecture, policies, and practices of the healthcare provider.

The HISQ is designed as a questionnaire, not a survey, and it is expected to be filled out in its entirety just once, but collaboratively, using the appropriate technical and leadership resources from across the organization. It is also recommended that the questionnaire be completed through a series of iterative drafts whereby there are active discussions about both the questions and answers. This will ensure there is good understanding of both the question be asked and the response given.

Information Technology Architecture Review (ITAR)

In addition to the completing the HISQ, the organization submits to a full examination of their IT architecture. This review involves obtaining network diagrams, data center diagrams, network device configurations, and other documents that depict how the network and infrastructure architecture is implemented. The topology of the environment is scrutinized specifically for appropriate isolation and segregation of ePHI data on the organization's network.

The HIPAA regulations specifically address transmission security in §164.312(e) (1) by the following statute, "implement technical security measures to guard against unauthorized access to electronic protected health information that is being transmitted over an electronic communications network" (Title 45-Public Welfare, 1996). The regulation goes on to state that there are 2 key components of ensuring the security of ePHI during transit: integrity controls and encryption. The primary purpose of integrity controls is to ensure the ePHI data isn't modified in any way during transmission. Encryption serves to disguise the true content of data such that it is not easily readable or decrypted without proper authorization. These 2 security measures are the basic foundation of providing secure transmissions (Acharya, Coats, Saluja, and Fuller, 2013). If an unauthorized entity can't read the contents of a transmission or alter or delete any portion of it, the authenticity and confidentiality of the transmission is ensured. While the concepts are straightforward, successfully achieving them can be challenging.

There are a number of fundamental approaches that are effective across almost all environments. It is important to acknowledge that before making architectural decisions related to the technical aspects of transmission security, it is imperative that operational needs, functional and financial, be considered. It is easy for the technical staff typically tasked with the implementation of the HIPAA technical safeguards to lose sight of how the technology will actually be used in practice. If the chosen measures provide the appropriate levels of security but are impractical to utilize, the overall solution is ineffective. Further, in such cases the likelihood of both intentional and accidental misuse or circumvention of the organization's security will increase dramatically. The ITAR performs a thorough analysis of the IT architecture and provides an evaluation using the following considerations: Usability, Security and Dependability.

- **Usability:** The verification is to ensure if the systems are functional as needed for normal business operations and if the users can reasonably reach the data, they need from the operational locations.
- **Security:** The verification is to ensure that the entire ePHI data is appropriately isolated and segregated on the network.
- **Dependability:** The verification to check for single points of failure within the architecture that will adversely affect business continuity in a disaster recovery situation.

Healthcare Practitioner Survey (HPS)

The last assessment in Phase 1 is the Healthcare Practitioner Survey. This assessment evaluates the organization's human-technology interaction by the healthcare practitioners. The survey covers the healthcare personnel's perception of the current IT practices, their understanding of requirements and procedures in place, and their specific interactions with ePHI data. It is not uncommon for an organization's published and intended IT security practices to not directly correspond to how their users are actually functioning (Johnson, Goetz and Pfleeger, 2009). This assessment's purpose is to provide a check and balance for established policy and procedures that were examined in the HISQ. The survey is designed to be short but engaging, consisting of 25 'yes-no' questions related to the practitioners' awareness of the healthcare provider's IT policies and practices. The survey should be presented electronically and completed anonymously to encourage honesty and frankness. Once the survey has been completed, the results are compiled and evaluated. The findings of each of the assessments are combined to produce a cumulative Phase 1 summary, presented as the Comprehensive Organization Assessment and Roadmap (COAR). After creating the COAR, Phase 2 performs a practical evaluation of the areas covered in the first phase and amends and expands the COAR as necessary.

Comprehensive Organization Assessment and Roadmap (COAR)

The COAR is effectively the framework's master report of the results of both Phase 1 and Phase 2. At the conclusion of Phase 1, an initial draft of the COAR is produced that contains the results of the all the Phase 1 assessments, along with any recommended mitigating actions. A thorough organization review of the COAR is very useful at this stage, prior to beginning Phase 2. Each question of each questionnaire and survey for all Phase 1 assessments contains a cross-reference to both the HIPAA statute and the corresponding section of the HISG. As such, the recommendations from the Phase 1 assessments can be easily combined with the guidelines laid out in the HISG, to produce a clear set of actionable tasks. Phase 2 shifts the assessment style from abstract to practical. Following the completion of Phase 2, the COAR will be revised to include the results from those assessments as well. Once the results of Phase 2 are included, the COAR will serve as a detailed implementation guide for the organization to follow in order to achieve HIPAA compliance.

Healthcare Information Security Testing Directive (HISTD)

When considering an evaluation of information security, an organization must first establish what the actual objectives are for the environment being examined. After the security objectives have been established, the actual test plan or methodology can be drafted. It is important to recognize that an effective testing plan must be easily repeatable. It is in the repetition of the security tests and surveys that many issues can be identified using comparative analysis of prior test results. Many times, issues or vulnerabilities are not immediately obvious during the course of normal examination but when compared to prior test results, anomalous conditions can be much more readily recognized.

The proposed security testing plan, the HISTD, divides the testing techniques into five key areas: target identification and analysis; target vulnerability validation; password cracking; business process testing, and application assessments. The identification and analysis testing are centered on network discovery, port and service identification, and vulnerability scanning. The vulnerability validation category consists of a variety of penetration tests on the different components of the organization information technology environment. The password cracking area is focused very specifically on testing the strength of passwords within the organization. The business process testing portion, much like the Healthcare Practitioner Survey, provides an examination of how technology is actually being used in normal business operations to ensure security controls are not being circumvented in actual practices. The final testing technique of application assessments is intended to provide in-depth application security testing beyond typical penetration testing.

Unfortunately, no single security test can be used to validate all systems and services from all perspectives. As such, it is necessary to use an assortment of tools to achieve a truly complete assessment. This research has focused on creating a collection of testing tools that can provide a comprehensive set of tests with the minimal amount of overlap. The collection of tools configured and preloaded on the two Tester Virtual Machines (VMs) are depicted in Figure 6. Additionally, the tests have been preconfigured and automated as much as possible to minimize the amount of effort necessary to conduct the testing.

Since security testing is a very fluid and changing process, it is recommended that all organizations establish an information security testing environment to become acquainted with the testing tools and run simulated tests to perfect the organization's testing plan. Figure 7 depicts a basic testing environment that was created by this

Figure 6. Security testing configurations

VM 1	
Operating System - Ubuntu 11.10	
Testing Tools	
Nessus 5.0	*Vulnerability Scanning*
VM 2	
Operating System - BackTrack 5 R3	
Testing Tools	
NMap	*Network Enumeration and Port Scanning*
THC-AMap	*Protocol Detection*
Enum4Linux	*Windows Enumeration*
Swaks	*SMTP Testing*
SSLScan	*Encryption Testing*
Bluediving	*Bluetooth Penetration Testing*
AirCrack	*Wireless Penetration Testing*
SMAP	*SIP Scanning for VoIP*
OneSixtyONe	*SNMP Scanning*
SQLMAP	*SQL Injection and Database Takeover Testing*
Armitage	*Exploitation testing*
THC-Hydra	*Password Cracking*
W3af	*Exploit testing*
Uniscan	*Website Vulnerability Scanning*
Nikto	*Web Application Testing (White box, Black box)*
Burpsuite	*Web Application Testing (White box, Black box)*

research and can be utilized by any healthcare organization. Having a dedicated testing sandbox environment can be helpful to show how each type of test is performed and understand their impact to the systems being tested. It is important to perform all types of security testing from both an internal and external perspective. In order to truly validate adequate security exists within the environment the conditions of the tests must match or be relevantly comparable to the scenario being tested.

Target Identification and Analysis

For the target identification and analysis tests, the systems on the network segment being tested will be cataloged including each system's operating system (OS) information and patching status as well as any open ports or active services. Network discovery can be performed using either an active scanning tool or passively using a network sniffer. While the passive approach tends to make the lesser impact to the performance of the network or scanned machine, it takes considerably longer, and the results are bound by what events are actually taking place. Active scanning usually yields but more comprehensive results and allows the scans to be targeted to look for specific characteristics, regardless of a system's current activity.

Figure 7. Security testing environment

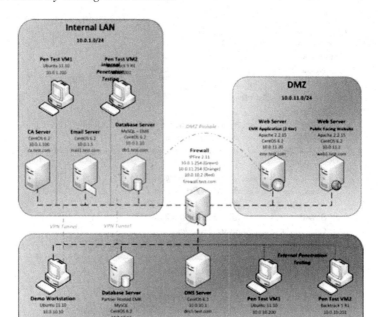

It is important to recognize that discovery scanning can be an intensive process and potentially have significant impact on the systems it is scanning and in cases of older systems, cause system failures. Network discovery can be helpful to detect unauthorized systems present on an organization's network. It is important to note that scanning should not be limited just to the wired network. A number of wireless scanners exist that can very effectively collect relevant data about wireless devices and the local wireless network that wouldn't normally apply to a traditional wired network. The wireless scanning should not only include all 802.11 channels but also Bluetooth and a general radio frequency (RF) spectrum analyzer.

Once the connected systems have been identified for a particular network segment, these hosts are further examined using a port scanner to see which ports open and what services are are running on those ports. The port scanning process can also perform OS fingerprinting. The last test of this group is to perform vulnerability scanning. The types of checks vulnerability scanners can perform depends on the tool, but they typically can identify out of date software, missing patches, and various errors with configurations. Unfortunately, vulnerability scanning has a number of limitations that are important to recognize. First, vulnerability scanning is much like virus scanning as it relies on a repository of signatures and therefore can only detect documented issues. This requires frequent updating of the repository to be able to discover the latest vulnerabilities. Secondly, these scans usually have a high false positive error rate and thus require an experienced information security individual to effectively interpret the results. These weaknesses ultimately limit the scanning process as there are considerable portions of these tests that are labor intensive and cannot be automated. The network-based scans generate significantly more network traffic that network discovery and port/service scanning and can prove harmful to the hosts being scanned. When the vulnerability scanning is complete, the tests in first stage of the security testing plan will have produced a comprehensive report of the organization's connected systems and including information about their OS, active services, and any vulnerability they have.

Penetration Testing

With the information generated in the first round of the testing, the next stage will continue to search for vulnerabilities and demonstrate the exposures created when they are exploited. Penetration testing will simulate real-world attacks and provide information about how the system, application, or network will respond to malicious attacks. Penetration testing also can help provide information about effective countermeasures to attacks, how to detect an attack, and the appropriate response. Penetration testing is very labor intensive, much like vulnerability scanning, and as such typically requires a professional with considerable skill to conduct the testing successfully without damaging the targeted system. The majority of the tests performed as part of this security testing framework fall into the penetration testing category.

Password Cracking

After the penetration testing stage is complete, a series of password cracking tests are performed. There are a couple general approaches for password attacks: dictionary attacks, brute force, and rainbow table attacks. Typically, password cracking involves obtaining the hash of the actual password, either from the target system directly or using a network sniffer. Once the hash is obtained, the attacks take different approaches in an attempt to generate a matching hash to discover the actual password. While an attack could be directed at a system service or application, these attacks are typically not as efficient and take considerably longer to conduct. Using this approach, you are limited to response times of the target system or application per attempt, as well as the round-trip network time, to determine if the attack was successful. While the time associated with a single attack is extremely small, when millions of credentials are being attempted, the compounded time usually makes this approach unattractive. By having a copy of the hash, you are trying to recreate, the attack is only limited to the processing capabilities of the system performing the attack. Different from penetration testing, password cracking can be effectively performed offline to remove the possibility of any impact on the target system, network, or application. The objective of password cracking is to determine how predisposed an organization's password policies are being compromised. In cases where passwords are determined to be vulnerable, their respective strength can be augmented to achieve appropriate entropy.

Business Process Testing

While examining each component of an IT environment is a critical exercise, it is also important to examine entire processes to verify each component is being used appropriately during normal business practices. It is possible that not all security capabilities of each component are actually employed in practice or exceptions have been 'built-in' to processes that circumvent the safeguards the components could normally exert. Clearly, EHR security is a crucial element of any healthcare organization's overall security framework. To this end, the ways in which EHR systems are used in normal practice serve as excellent candidates for process testing scenarios. From a process perspective, EHR security can be examined in three key areas: access, transmission, and storage.

- **Access:** This category deals specifically with the functional areas related to authentication, authorization, and delegation. More specifically this area handles who can have access to data, which data they can access, what type of actions they can take on that data, and who they can provide some degree of access to the data as well.
- **Transmission:** This category covers how data is moved in an electronic medium. This area covers where data can be access from, where that data can be sent to, how the data is formatted while being moved, how data is presented to the user, and what mechanisms can be used to send the data.
- **Storage:** This category accounts for how data is captured and preserved. This area deals with how data can be added, modified, or deleted, how the data is validated upon entry, the format of how data is stored electronically, how the data is preserved, and how data integrity is ensured.

Application Review and Testing

This part of the HISTD involves an extensive review, categorization, and analysis of all enterprise applications. Each application is examined to determine if it interacts with ePHI and if so, in what way and for what function or purpose. This final type of testing is directed specifically at an organization's applications that capture, access, or transmit ePHI. This type of testing involves both *white box* and *black box* approaches. White box testing takes the perspective of an internal user such that the tests assume a working knowledge of how the application works. Conversely, Black box testing assumes the attacker has no familiarity with application or how it is designed and implemented. These types of tests and attacks include injection attacks, file descriptor attacks, data corruption attempts, and intentional misuse of the application beyond the organization's published policies and procedures. Application testing, along with all the other parts of the vulnerability validation-testing phase are used to evaluate systems during actual use. Therefore, the closer the tests are to normal conditions, the more useful the results of the tests will be in discovering potential risks.

Phase 2: Information Security Testing

Phase 2 is a detailed, hands-on technical review and assessment of the IT environment. This phase measures and analyzes the actual performance of the systems and practices both against the theoretical goal of the HISG and the reported state of the organization provided in the assessment stage of Phase 1. The variances found in this effort are reflected in the COAR with appropriate mitigating actions. The technical review includes onsite visits, penetration and vulnerability testing, and a comprehensive review and assessment of all enterprise applications.

Interviews and Inspections

The interviews and inspections stage of Phase 2 is aimed at providing an opportunity to inspect various components of the IT environment including physical security controls for the data center and other locations where ePHI data is stored. While this was evaluated in Phase 1, these inspections should serve as the effective penetration tests of the physical computing environment. The onsite visits should involve interviews with all appropriate personnel of the organization, both within the IT department, and administration, and leadership.

Penetration and Vulnerability Testing

In addition to the onsite visits, the IT staff is engaged to conduct penetration and vulnerability testing on the network and infrastructure portions of the organization. All associated testing is documented in the Healthcare Information Security Testing Directive (HISTD). The HISTD ensures the testing is standardized and easily repeated not only during the current review period but in the future as part of the organization's continued compliance efforts. This stage will simulate real-world attacks and provide information about how the system, application, or network will respond to malicious attacks. The penetration and vulnerability testing also can help provide information about effective countermeasures to attacks, how to detect an attack, and the appropriate response. Business process testing is an important aspect of this stage. This aspect examines entire processes to verify each technological component is being used appropriately during normal business practices. Many information security breaches are actually caused by a failure to use a system as designed or the procedure doesn't match the policy.

EHR Security and Privacy Assessment

The last task of Phase 2 is to perform an in-depth review of the organization's EHR systems specifically. This assessment examines both the security and privacy policies and practices. The evaluation instrument is a survey that is completed by the leadership responsible for the technical support of the EHR system. The survey is broken into 3 main sections - organization policies and practices, functional implementation, and technical implementation. The first part, organizational policies and practices, covers topics such as how staff is trained on HIPAA privacy requirements, security awareness training, the presence and application of acceptable use policies, how ePHI releases are handled, and how data alteration/deletion is guarded against. The functional implementation section covers how the EHR system is used in normal operations. Questions for this section cover how the business practices for how ePHI is captured, accessed, and transmitted. The last area of the survey, technical implementation, examines how the EHR system was deployed technically including the system architecture, how patch management is addressed, presence of intrusion detection and prevention, and finally network location and safeguards. The information captured within this survey provides a complete portrayal of whether the organization has enacted adequate security and privacy controls for their EHR systems. Once each of the technical reviews is complete, the final task of this phase is to update the COAR report with all the findings and corrective actions identified in this phase. At the conclusion of this phase, the organization's entire IT environment has been methodically examined and evaluated.

Phase 3 – Implementation

The final phase involves taking the findings of the first two phases and performing corrective actions as appropriate. Phase 3 is the implementation stage including changes related to technical configurations, policy, procedures, training, and documentation. At the start of the implementation phase, an implementation plan will be drafted, based off of the final COAR. While the findings and recommendations laid out in the COAR will provide specific tasks to complete, a plan needs to be developed of how to put those changes into operation. Meetings with stakeholders, IT staff, and administrative staff will be necessary to create an effective plan including an appropriate timeline. Once the plan has been developed, the actual implementation can be scheduled and started. In addition to the technical, policy, and procedural changes covered in the COAR implementation plan, this phase will also ensure that necessary documentation is created for both the impending changes and the preexisting environment. Further, this phase will include any necessary training – administrative, technical, or functional – related to the changes implemented, new procedures, and general security awareness training of the organization moving forward.

Post-Compliance

With the completion of the third phase, the entire framework will likewise be completed. The designed result of the framework will first and foremost be the achievement of HIPAA compliance for the organization. In the efforts to attain compliance, there will also be the potential for a number of other tangible accomplishments. This framework will create a standardized Healthcare Information Security Guide that can be referenced and updated for perpetuity. The HISG will serve as a critical resource for evaluating future enhancements and changes to the environment and ensure compliance is maintained. Additionally, the framework will produce a series of valuable tools for periodic testing of the security configurations. These tools will provide important actionable information as well as save time and effort in regard to the ongoing penetration and vulnerability testing procedures. Lastly, this framework will afford extremely useful training and awareness of security to the organization at all levels. The assessment exercises alone will orient the healthcare providers, technical staff and administration alike on the current updated state of their IT environment. It is often the case in HIPAA compliance efforts, that the simple lack of knowing how to measure compliance can greatly delay the entire effort. This research educates organizations as to what compliance requires, how these requirements translate into their specific environment, and how to satisfy them quickly, efficiently, and at a significantly reduced cost compared to tackling this effort alone.

Evaluation

In order to validate the effectiveness of this research, it was vital that both frameworks be implemented in an actual healthcare provider's environment. This research was fortunate to have cooperative agreements with 2 national healthcare providers to provide that opportunity. The large central Pennsylvania hospital was engaged for evaluation of the Healthcare Information Security Compliance Framework (HISCF). Both entities are national hospitals with the PA hospital having over 500 licensed beds and more than 400,000 patient admissions (combined inpatient and outpatient) every year, while the Maryland healthcare provider has over 800 beds and more than

350,000 patient admissions (combined inpatient and outpatient) each year. Each of these hospitals interact with a significant number of patients annually and are both faced with the daunting and costly challenges of achieving Health Insurance Portability and Accountability Act (HIPAA) compliance and providing patient access to electronic health records.

Case Study of Healthcare Information Security Compliance Framework

Since the HISCF was largely borne out of a federal grant of which a key deliverable was compliance assessment, the PA hospital was very eager to participate in its implementation even though they had already obtained certification as a HIMSS Stage 6 Hospital. This partnership between the Pennsylvania hospital and Towson University started in 2011 and promised the hospital would be provided a comprehensive assessment of their entire IT environment, including specific, actionable tasks to remedy any deficiencies uncovered. The partnership was scoped for a 3-year engagement, with roughly 1 year allocated per phase of the larger information technology assessment framework. The HISCF, depicted in Figure 3, is designed to take an organization from the initial recognition of the need for Health Insurance Portability and Accountability Act (HIPAA) compliance all the way through to implementation of any necessary changes to their environment. Along the way to compliance, specifically in Phase 2, a comprehensive security audit is performed that partially satisfies the necessary attestation for Meaningful Use.

HISCF: Phase 1

Starting with Phase 1, a high-level assessment, involving a thorough review of all technology practices and architectural designs, was performed. The information technology staff was engaged to assist in the completion of both the HISQ and ITAR.

Healthcare Information Security Questionnaire (HISQ) Execution

The HISQ was presented to a single point of contact in the Pennsylvania hospital's IT group. This individual, a senior security engineer, then worked with the appropriate staff within the 52-member IT department to complete each part of the questionnaire. Once the initial draft of the HISQ responses was completed, a series of interviews were conducted to review the responses for clarity and consistency. The responses were also reviewed by the hospital's Chief Information Officer (CIO) for additional validation. The measurement scale used to quantify the responses is based on the percentage the organization is in compliance with the guidelines laid out in the

HISG with is directed based on the HIPAA guidelines and National Institute for Standards and Technology (NIST) recommendations or HIPAA implementations.

HISQ: Policy Assessment Findings

The policy and procedure review results for each of the 4 policy areas had a number of similarities that cut across many of the technology areas of the organization. The common theme was that the healthcare provider had addressed most of the needed areas to some degree but not completely, seemed to emerge very quickly from the results.

Disaster Recovery and Business Continuity

In the area of Disaster Recovery and Business Continuity, the organization had only partially implemented a DRP plan and the portions that did exist need significant updating. The hospital had been rapidly growing over the last few years and their patient counts had been equally increasing. As such, disaster recovery and business continuity planning had not been given the appropriate degree of attention. Based on the assessment findings shown in Figure 8, the organization's overall compliance with disaster recovery and business continuity rated 77% adequate. As expected, some areas were more complete than others. The organization scored over 80% in compliance for their recovery plan, their data and application criticality plan, their data backup and retention plan, and testing and revision process. However, the organization DRP policy, their emergency mode plan, DRP training, and documentation were all below 70% compliance with training only rating 50%. Some of the specific key findings in this area included the lack of backup copies of data being kept at an off-site facility. There was also no documentation for DRP training and the training that did exist was pretty limited. Another issue uncovered was that fact that there are single points of failure within both the recovery and emergency mode plans. Specific key responsibilities had no delegation accommodations therefore if a specific person is

Figure 8. Disaster recovery & business continuity outcomes

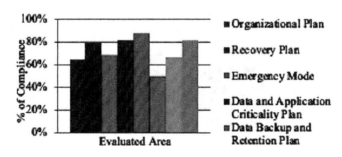

not available, those responsibilities and functions cannot be performed. This was a significant flaw in the organization's current DRP procedures.

Risk Management

The results for Risk Management shown in Figure 9 were considerably worse than the other 3 areas. Overall the organization rated just 52% in compliance. The ongoing risk management activities were by far the least adequate area, scoring just 18%.

The hospital had very little proactive risk monitoring in place. Most risk mitigation efforts were reactive once an issue has been uncovered. Similar to DRP, the organization had partially developed plans for risk analysis and assessment as well as mitigation. Unfortunately, none of these programs were fully implemented nor were they comprehensive enough to be in compliance all HIPAA guidelines.

Operations Management

The healthcare provider had considerably more comprehensive policies and procedures related to operations management as shown in Figure 10. The antivirus program was 100% compliant and both media handling and data disposal had only very minor deficiencies. Security monitoring was in fact the only aspect of this area

Figure 9. Risk management outcomes

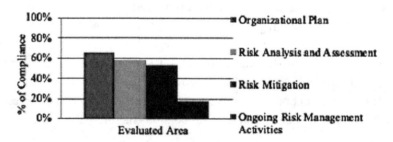

Figure 10. Operations management outcomes

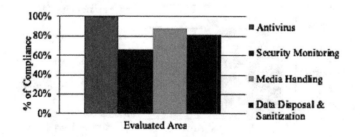

that had inadequacies of any significant degree. One of the main factors creating the issues related to security monitoring was that while they had a commercial intrusion detection and prevention system (IDPS) implemented, it had only been configured to monitor a very small segment of the organization's environment. Once the IDPS was fully configured this aspect should have come into compliance.

Logical Access

Logical access, shown in Figure 11, was measured at being 80% compliant overall. As with most areas assessed, plans for the various aspects of logical access had been developed and implemented but they were not comprehensive and were not up to date. It was discovered that many of the hospital's practices were not reflected in the policy nor were all the procedures mentioned in the policies actually in practice. Another key finding was that data could not be easily shared with external entities. While security of this data was sufficient, the logical access practices being employed created usability barriers and deficiencies. Further, due to the inflexible logical access issues, access to ePHI was not possible remotely. This situation also created an issue related to emergency access for business continuity during a disaster scenario.

HISQ: Technical Assessment Findings

In addition to reviewing the policies and practices of the organization, the HISQ is designed to perform a technical assessment of information security. This portion of the questionnaire is divided into 4 main IT areas: Network, Applications, Database, and Infrastructure. The relevant members of the healthcare's organization completed the questionnaire and those responses are denoted below. Only potential actionable issues were mentioned in this assessment. For any portion of the IT environment that is managed or hosted by a third-party, the assumption was made that those aspects of the environment were implicitly satisfactory and in compliance with all HIPAA regulations. A review of all third-party systems and management practices

Figure 11. Logical access outcomes

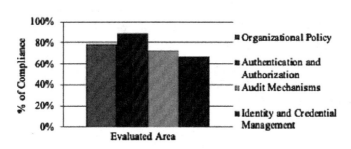

is out of scope for this assessment. For those potential issues that were identified, a description of the finding and the recommended corrective action to mitigate or remove the issues are provided. Regardless of whether a specific finding was cited with an actionable mitigating response, each area was scored based on how thoroughly and effectively it is being addressed within the current environment. The scoring determination was made from the series of responses given to each question, supplemental comments made in the Healthcare Practitioner Survey, and the general understanding of the environment based on all information provided.

Network

- **Policies and Procedures:** The response to the questionnaire indicated policies and procedures related to network operations and disaster recovery had not been adequately addressed. There seemed to be an indication that policies and procedures were once in place, but they were no longer up to date with the current environment. Having comprehensive policies and procedures established, documented, and published for review would have addressed a number of HIPAA guidelines – §§164.308(a)(1), 164.308(a) (2), 164.308(a)(5), 164.308(a)(7) – related to the security management process, assigned security responsibilities, security awareness and training, and contingency planning. The questionnaire response indicated that acceptable use polices (AUP) were in existence but were not up to date. It is important to have documented and published acceptable use policies that all users agree to and this consent is recorded. Having an AUP in place further satisfies HIPAA guidelines – §§164.308(a) (1), 164.308(a) (5) – related to security management process as well as security awareness and training. The questionnaire response indicated that staff members were regularly receiving security alerts and advisories and likewise took the appropriate actions. That practice in part satisfies HIPAA guidelines related to security awareness and training - §164.308(a) (5).
- **Practices:** The response to the questionnaire indicated that administrative credentials for network devices and applications were in some cases shared and, in some cases, unique to the staff member. The sharing of credentials, especially for administrative accounts with elevated privileges, is strongly discouraged. Without unique credentials the audit logging effectiveness is greatly reduced with respect to identity. The HIPAA technical safeguard guidelines related to access control – §164.312(a) (1) – clearly requires unique identifiers and credentials.

Since non-institutionally owned devices and PCs were allowed to access both the internal wired and wireless networks, it was critical that those machines had been scanned for risks prior to allowing full access to the network resources. Network access protection (NAP) and network admission control (NAC) software can provide the necessary safeguards required for security awareness and training – §164.308(a) (5). The response indicated that software to perform NAP/NAC was licensed but not implemented. The response further indicated that vulnerability scanning was performed but infrequently. It was crucial to regularly check the network for vulnerabilities and problems, and then address them in order to minimize the opportunity for accidental or malicious exploitation. The transmission security and regular evaluation requirements of HIPAA – §§164.308(a) (8), 164.312(e) (1) – necessitate regular penetration and vulnerability testing.

Database

- **Policies and Procedures:** The response to the questionnaire indicated policies and procedures related to database operations and disaster recovery had not been adequately addressed. The responses indicated policies and procedures had either not been created or had not been completed. Having comprehensive policies and procedures established, documented, and published for review would have addressed a number of HIPAA guidelines – §§164.308(a) (1), 164.308(a)(2), 164.308(a)(5), 164.308(a)(7) – related to the security management process, assigned security responsibilities, security awareness and training, and contingency planning. The questionnaire response indicated that staff members were regularly receiving security alerts and advisories but were not taking the appropriate actions. Responding to security notices is a required practice to satisfy in part HIPAA guidelines related to security awareness and training - §164.308(a) (5).

- **Architecture:** The response to the questionnaire indicated in some cases the database, application server, and web server all resided on the same physical machine. ePHI related applications that utilize databases should have use either a 2- or 3-tier architecture such that the database does not reside on the same physical server as either the application or web server. All efforts should be made to minimize the exposure each server has to non-administrative users and networks. Since all applications must inherently be accessed by internal and/or external users, some amount of exposure is necessary. Through a 2 or 3-tier architecture, the only way users can access ePHI data is via proxy through the web or application servers. This minimizes the impact of a breach at a web server since no actual ePHI data resides on those servers.

The response further indicated that there was no redundancy for databases within the environment. Redundancy should exist for databases as appropriate to the sensitivity or criticality of the data they hold. Having redundancy for databases that hold ePHI will assist in providing business continuity and satisfy the HIPAA regulations related to contingency planning – §164.308(a)(7).

Data encryption is critical for protection of ePHI. The response to the questionnaire denoted that encryption occurred only in some cases. In order to satisfy the HIPAA technical safeguards related to access control – §164.312(a) (1) – all ePHI data must be encrypted while at rest.

- **Practices:** The response to the questionnaire indicated there was no monitoring or alert mechanism in place for databases. The indication was that activity logging was in place and was referenced reactively as required when an issue occurred. Having monitoring for database activity is crucial to ensuring the security of ePHI data and knowing what is happening within the database. The monitoring should have the ability to alert the appropriate staff as well as either automatically or manually responding to events. HIPAA regulations – §164.312(b) – require adequate audit controls be in place for both non-repudiation and exception notification.

Administrative credentials for databases were shared among staff members according to the response. The sharing of credentials, especially for administrative accounts with elevated privileges, was strongly discouraged. Without unique credentials the audit logging effectiveness was greatly reduced with respect to identity. The HIPAA technical safeguard guidelines related to access control – §164.312(a) (1) – clearly requires unique identifiers and credentials.

Applications

- **Policies and Procedures:** The response to the questionnaire indicated policies and procedures related to application operations and disaster recovery had not been adequately addressed. The response indicated policies and procedures had either not been created or had not been completed. Having comprehensive policies and procedures established, documented, and published for review would have addressed a number of HIPAA guidelines – §§164.308(a) (1), 164.308(a)(2), 164.308(a)(5), 164.308(a)(7) – related to the security management process, assigned security responsibilities, security awareness and training, and contingency planning. The questionnaire response indicated that staff members were not regularly receiving security alerts and advisories and likewise were not taking the appropriate actions. Distributing and responding to security notices is a required practice to satisfy in part HIPAA guidelines related to security awareness and training - §164.308(a)(5).

- **Functionality:** The response to the questionnaire indicated ePHI related applications had audit logging capabilities that produce easily reviewable logs. However, it was stated that the logs were not easily searchable and there was no central management of these logs. Audit logs that are centrally managed and searchable enable monitoring and alert functionality for proactive security. Adequate audit controls are a HIPAA requirement – §164.312(b) – including the ability to review and search exception reports.

The response further indicated that encryption was not used when ePHI data was transmitted between applications. Encryption is an effective way to safeguard the integrity of data while at rest and in transit. All methods used to transmit ePHI data between applications or within the application itself should use secure channels and some form of encryption. HIPAA regulations related to access control and transmission security – §§164.312(a) (1), 164.312(e)(1) – require encryption to be used when reasonable and appropriate.

All ePHI related applications should have the ability to check their data for accuracy, completeness, and validity. The response indicated that not every relevant application had this capability and furthermore SQL injection vulnerabilities had been identified for some applications. Invalid data can create both intentional and unintentional data pollution. Application and/or database level data checks should be used to mitigate the risk of compromised data integrity and address HIPAA regulations related to audit controls and integrity – §§164.312(b), 164.312(c)(1). The possible methods of ePHI data extraction and transmission were not readily known according to the questionnaire response. It is a fundamental HIPAA requirement – §164.308(a) (8) to have an accurate understanding of how ePHI can be accessed and moved within the electronic environment of an organization. Without an adequate understanding of how users and applications interact with ePHI data it is impossible to take sufficiently secure measures to safeguard said data. All ePHI relevant applications should have all possible methods of ePHI data extraction or transmission secured and documented including aggregations of ePHI data outside of enterprise applications and databases.

- **Practices:** The response to the questionnaire indicated that it was unknown whether administrative credentials for applications were shared or unique to the staff member. The sharing of credentials, especially for administrative accounts with elevated privileges, was strongly discouraged. Without unique credentials the audit logging effectiveness is greatly reduced with respect to identity. The HIPAA technical safeguard guidelines related to access control – §164.312(a) (1) – clearly requires unique identifiers and credentials.

Infrastructure

- **Policies and Procedures:** The response to the questionnaire indicated policies and procedures related to infrastructure disaster recovery had not been adequately addressed. The response indicates policies and procedures had either not been created or had not been completed. Having comprehensive policies and procedures established, documented, and published for disaster recovery would have addressed HIPAA contingency planning – §164.308(a) (7). The questionnaire response also indicated that an accurate inventory of all institutional hardware had not been created or was incomplete. Having a complete, accurate inventory of the organization's hardware will address in part the HIPAA regulations related to workstation use and security and device and media controls – §§ 164.310(a)(1), 164.310(b), 164.310(c).
- **Architecture:** The response to the questionnaire indicated that there was redundancy for servers in some cases but not for all servers within the environment. Redundancy should exist for all servers as appropriate to the sensitivity or criticality of the data they hold, they interact with, or transmit. Having redundancy for servers and therefore the services or applications they hold will assist in providing business continuity and satisfy the HIPAA regulations related to contingency planning – §164.308(a)(7).

The response also indicated that servers were not located on segregated networks from both external hosts and internal user workstations. Network segmentation in conjunction with 2 or 3 tier application architecture allow for greater security through minimizing exposure. Managing information access and exposure is a HIPAA requirement – §164.308(a) (4). No intrusion detection/prevention systems (IDS/IPS) was in place in the environment according to the responses. IDS and IPS provide many tools and techniques to monitor and react to intrusion events, detect and mitigate attacks, and provide notification of unauthorized system use. Most operating systems have some degree of IDS capabilities built-in but may need to be configured and enabled to provide the functionality. An effective IDS/IPS strategy utilizes both the delivered capabilities of the operating systems as well as a stand-alone IDS/IPS application. Monitoring the servers and workstations of potential intrusions both electronic and physical is a requirement of providing adequate security – §164.310(a) (1).

- **Practices:** The response to the questionnaire indicated that unregistered devices/machines were permitted to use NOS resources such as file and print sharing. This type of access implied NOS resources allow anonymous access which was not a secure practice. HIPAA regulations related to workstation security - §164.310(c) – require methods of access to be documented. Anonymous access greatly complicates the accurate recording of access activity.

The response further indicated that users had the ability to modify their PC/device configurations as well as install additional software. In such cases, it is important that users be trained on appropriate security best practices to help guard against unintentional compromises through the installation of malware or other hostile applications. For PCs and devices that have access to ePHI data, users' ability to modify the configuration and install software should be limited as operationally practical. The greater the capacity for users to modify their workstations increases the risk for compromise and likewise must be addressed as part of HIPAA regulations for workstation security – §164.310(c).

There were no measures in place to address ePHI data loss in the event a PC or mobile device was lost or stolen according to the response. At a minimum file encryption and strong device authentication should have been used to safeguard ePHI data if the device it was stored on was no longer in possession or control of the user originally authorized to access it. Many mobile devices have the ability to complete delete their contents remotely if they are attempted to be broken into with brute force or other attacks. Data loss prevention (DLP) measures satisfy in part the HIPAA regulations related to device and media controls – §164.310(d) (1).

Server hardening is an industry best practice that was not being performed according to the response. Hardening ensures only the minimally necessary access and exposure for a server and the services or applications that it hosts. Many malicious attacks exploit unused, accessible resources on servers to compromise those systems.

Information Technology Architecture Review (ITAR) Findings

A number of phone interviews and exchange of emails were performed to gather information about the organization's IT architecture. Network diagrams, system configuration documents, and hardware specifications were examined as part of the architecture review. In contrast to the HISQ where the IT staff answered questions about the organization's policies or technical implementation decisions, the ITAR and subsequent analysis was performed by the team at Towson. Certainly, in subsequent reassessments, the organization could perform this step themselves. The ITAR revealed there were a number of critical areas not addressed in the network design. The network topology was analyzed in detail and determined to be flat in crucial areas which indicates redundancy was not present in all areas. Further, network paths were not optimally designed for enhanced performance. The review also pinpointed a number of single points of failure within the network design thereby not sufficiently satisfying the HIPAA guidelines for contingency planning and business continuity – §164.308(a) (7).

It was documented that VLAN segmentation was not present throughout the network. VLAN segmentation is an essential technique to securing communications within an organization. One part of network segmentation is creating a DMZ in which all publicly accessible web servers are located. According to the review interviews, a DMZ existed but was not effective. Furthermore, it was indicated that not all publicly accessible servers were located within the DMZ implying that portions of the internal network were reachable directly from external hosts. It was also indicated that internal VLANs were not always appropriately segregated from each other thereby enabling unnecessary accessibility to secure resources and data. VLAN segmentation is one aspect of ensuring systems and data is not unnecessarily accessible by internal and/or external hosts – §164.308(a) (4).

The review further determined that all applications and databases could have been accessed directly using a wireless connection. Wireless networks are inherently insecure due to the nature of the transmission medium and the inability to control where the transmission travels and therefore who can receive or intercept it. While there are measures possible to minimize wireless networks' vulnerabilities, they should be regarded as an insecure medium and only used for such applications and services that are tolerant to the intrinsic risk or required for operational necessity.

According to the interview responses, there was an absence of stand-alone intrusion detection/prevention systems (IDS/IPS) within the environment. IDS and IPS provide many tools and techniques to monitor and react to intrusion events, detect and mitigate attacks, and provide notification of unauthorized system use. Many network devices have some degree of IDS capabilities built-in but may need to be configured and enabled to provide the functionality. An effective IDS/IPS strategy utilizes both the delivered capabilities of the network devices as well as a stand-alone IDS/IPS application. Monitoring the network is a requirement of providing adequate transmission security – §164.312(e) (1).

Healthcare Practitioner Survey (HPS) Findings

The focus group used for this survey was approximately 400 healthcare staff from the Pennsylvania hospital and its partner clinical practices. The group's population is diverse in gender, race, ethnicity, and creed.

Figure 12. HPS - ePHI access outcomes

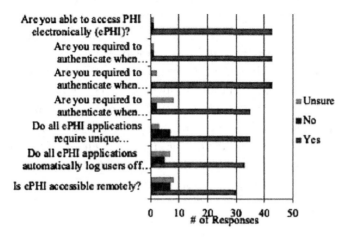

All members of the focus group were qualified physicians or physicians' assistants at the hospital or clinical practices and appropriately familiar with the policies and practices of the hospital. The survey was completed anonymously to ensure honest, accurate responses as well as remove any undue bias from the analysis of said responses. The survey had a little over 10% response rate, resulting in 45 total responses received as detailed in Figure 12.

The first 7 questions of the HPS were all relevant to the HIPAA Security Rule. The purpose of these questions was to provide a set of baseline questions to ensure the respondents did indeed work with ePHI and had a basic familiarity the hospital's computing environment. Over 95% confirmed this familiarity in questions 1-3 and this only dipped as low as 70% as the complexity of the questions increased about general accessibility of ePHI at the healthcare provider. It is significant that almost 20% of the respondents were unsure whether authentication was needed for imaging applications. Federal regulations require authentication for all access to any application that holds ePHI. The responses of this question suggest that the organization was meeting this requirement satisfactorily and a significant part of the population was unfamiliar and potentially uninvolved with imaging applications. It was also noteworthy that 16% of the survey responses stated that shared accounts were in use to some degree for ePHI-relevant applications. Federal regulations mandate user accounts for ePHI applications be unique per individual for auditing and non-repudiation. The responses suggest that the majority of ePHI applications were using unique user accounts but not all. It is critical for the healthcare provider to review the authentication model for each ePHI application and implement user-specific accounts for any application that doesn't already employ that scheme. Similarly, around 12% of the responses stated automatic log offs did not occur for all ePHI applications. HIPAA regulations clearly require all applications that interact with ePHI to automatically log users off after a period of inactivity. According to the responses, a comprehensive review of all ePHI relevant applications was needed to ensure each application had this capability enabled. There was a specific comment that some applications within the hospital kept the original user logged in indefinitely.

The next group of questions had to do with how ePHI is controlled, including how it can be replicated, where it can reside, and security around its transmission. Based on the range of responses, it was not universally clear whether ePHI data could be saved on mobile and/or personal devices or included in emails. The responses were somewhat split across the board as to the perceived or actual capability of taking ePHI data outside of the organization's data center or including it in email messages.

While the capacity to perform either activity is allowable within the HIPAA regulations, it becomes increasingly more challenging to maintain and demonstrate control of that data. Furthermore, if the organization does allow ePHI data to be included in and/or attached to emails, it is recommended that measures be taken to ensure its integrity. Digital signatures, encryption, and Data Loss Prevention systems are possible mechanisms that can be used for increasing the security of ePHI data included in email. As to the location of ePHI storage, about 74% of the respondents stated all ePHI data was stored within the organization's data center and none of the other responses contradicted the assertion. Control of all ePHI data is required to satisfy HIPAA regulations and having a common, centralized location to store all data makes the control of that data manageable. Similarly, the range of responses about how ePHI is captured suggests that there is not a clear, organizational understanding of all methods for capturing and storing ePHI data.

Figure 13HPS - ePHI data control outcomes

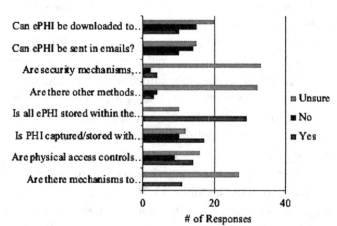

HIPAA regulations mandate that ePHI be stored in electronic format for interoperability with other healthcare providers and payers. 26% of the responses indicated that there were non-electronic methods being used and a number of additional comments expanded upon this assertion noting that there was considerable data storage using paper. One comment described the environment as half paper and half 'scanned' paper, which may in of itself not have been a completely accurate portrayal of the entire organization, but it did suggest improvements may have been necessary to achieve the electronic storage requirements.

Furthermore, the responses indicated that many locations that store ePHI data were secured physically but not all locations had been adequately addressed. With almost 23% of the respondents stating that physical controls were not present in all ePHI relevant locations that indicated some areas either had no or insufficient controls for ingress/egress. HIPAA regulations require physical access be secure and monitored. Any areas that did not have these controls in place had to be corrected.

Following the questions related the security of how ePHI was controlled, there were a section of questions related to the healthcare provider's privacy practices and policies as detailed in Figure 13 and Figure 14. Almost 77% of the respondents stated policies and procedures were in place for each of the proposed situations related to ePHI data releases and none of the other responses contradicted the assertion. Additionally, 72% of the respondents stated ePHI data releases were documented and securely recorded and none of the other responses contradicted the assertion. HIPAA regulations clearly require such policies and procedures to exist and require ePHI data releases to be documented and securely stored.

Figure 14. HPS - ePHI integrity & privacy outcomes

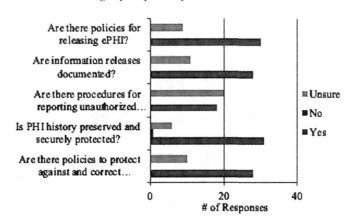

Based on the responses the indication was the organization was satisfactorily meeting this requirement. Similarly, about 82% of the respondents stated ePHI data history was preserved and protected and there was only 1 response contrary to the assertion. Nearly 74% of the respondents stated policies and procedures existed to address ePHI data being changed or deleted and there were no responses that contradicted the assertion. HIPAA regulations require history to be securely stored for all ePHI data and safeguards be in place to ensure the integrity of ePHI data to include any changes or deletions. Based on the responses the indication was the organization was satisfactorily meeting that requirement. Finally, just less than half of the responses stated that procedures existed for reporting unauthorized or inappropriate releases of ePHI data and no responses contradicted the assertion. The other half of the responses were unsure whether such procedures existed or not. HIPAA regulations mandate procedures be established to report and react to ePHI data being released unintentionally. While no responses indicated procedures didn't exist, the lack of understanding by the staff about such procedures in of itself created an implied deficiency.

The last group of questions related to the organization's policies and their staff's knowledge and awareness of those policies. Only a third of the respondents definitively stated that disaster recovery and emergency plans existed and were periodically tested. The majority, 63%, of the responses indicated that it was unclear whether plans existed or were tested. Business continuity is a HIPAA regulation of which disaster recovery and emergency plans are a critical component. Almost 78% of the respondents stated a HIPAA privacy training program existed and none of the other responses contradicted the assertion. Privacy training is a requirement of the HIPAA regulations and based on the responses the indication was the organization

Figure 15. HPS - policy outcomes

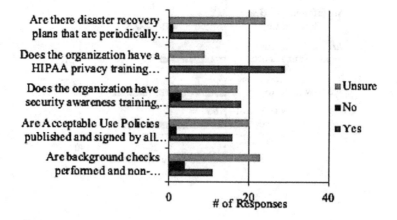

was satisfactorily meeting that requirement. Just under half of the respondents stated security awareness training was provided on a regular basis although almost as many responses indicated they were unsure if such training existed. Likewise, about 42% of the respondents stated acceptable use policies were published and signed by all users with access to ePHI data. However, almost 53% of the responses indicated they were unsure if such policies existed or were signed by ePHI relevant users. HIPAA regulations clearly require routine security awareness training and acceptable use policies be created, users agree to them, and this agreement must be documented prior to accessing ePHI data.

Based on the responses the indication was while privacy training and acceptable use policies existed within a portion of the organization, they were not pervasive through all units. Roughly 29% of the respondents stated background checks and non-disclosure agreements were signed prior to access being granted to ePHI data. About 11% indicated those practices were not present within the organization and 60%, the majority of responses, were unsure as depicted in Figure 15. HIPAA regulations require all personnel, business associates, and contractors to be adequately screened prior to gaining access to ePHI data. Based on the responses, the indication was that some units were performing the necessary screening, and some were not.

After all assessments were completed and reviewed, each area was rated based on the organization's degree of compliance. Compliance scores were provided for each section and sub-section to give indications where technical and organizational changes may be necessary. For each assessment, an initial draft, with any potential findings, was presented to the organization for their review and acceptance. The healthcare system either accepted the findings or disputed them and provided supporting documentation that demonstrates the finding was not valid. Following the review and acceptance process, the complete COAR report was produced and submitted to the organization for final review and acceptance.

Phase 1 Critical Findings

The three assessments in Phase 1 yielded a significant number of critical findings within the organization's environment. Considering the partner healthcare system is a HIMSS Analytics Stage 6 Hospital, the findings were non-trivial and representative of typical hospitals in the United States. The top critical findings based on organizational impact are detailed below – in decreasing order of criticality – along with their recommended corrective actions.

- **Single Points of Failure:** Analysis of the network topology determined the organization had six significant single points of failure related to how the various buildings on campus were connected both to the institution's data center and the Internet. In some of the cases those single points of failure were due to a single physical transmission medium existing between buildings.

The other cases were that not all buildings had redundant network paths to the internet or the data center. There were three instances where a disaster scenario in one building would segregate one or more buildings by extension from all other networks – internal or external. While a disaster scenario at a particular building is expected to directly impact that building's connectivity, such an impact should not be entirely debilitating to ancillary buildings. Single points of failure create an organizational risk to both contingency planning and business continuity, both of which are required – §164.308(a) (7) – within the HIPAA regulations. Redundancy within the network can be achieved using a variety of hardware and/or software solutions that was detailed in the COAR.

Disaster Recovery (DR) and Emergency Plans: Only a Third of the Healthcare

Practitioner Survey respondents definitively stated that disaster recovery and emergency plans existed and were periodically tested. The majority, 63%, of the responses indicated that it was unclear whether plans existed or were tested. This was reiterated by the responses provided to the HISQ by the organization's technical staff. The lack of awareness of a DR plan is akin to not having a plan altogether since the majority of personnel have not reviewed or tested the plan. Business continuity is a HIPAA regulation – §164.308(a) (7) – of which disaster recovery and emergency plans are a critical component. Disaster recovery plans must be established and periodically tested in order to be fully compliant.

- **Undue Exposure in Application Architecture:** It was discovered that not all applications that interact with ePHI data utilized a 2 or 3-tier architecture. Numerous applications were not configured such that web services, application services, and database services were segregated from one another. In many cases all these services resided on the same physical machine and were directly accessible by internal and external hosts. The final COAR recommended that all ePHI data that would be accessed by users should be done via a 2 or 3-tier application architecture with the data store on an internal, inaccessible network segment. All efforts should have been made to minimize the exposure each server has to non-administrative internal and external networks. Since all applications must inherently be accessed by internal and/or external users, some amount of exposure is necessary on generally accessible networks. Through a 2 or 3-tier architecture, the only way users can access ePHI data is via proxy through the web or application servers. This design minimizes the impact of a breach at a web server since no actual ePHI data resides on those servers.

- **Undue Exposure in Network Architecture:** The organization did not have an adequate demilitarized zone (DMZ) configuration that contained all publicly accessible web servers. Many of the application's web servers resided in the same network subnets where the application and database servers were located. In order to minimize exposure, any web server that is publicly accessible should reside in the DMZ and there should be no publicly accessible machines outside of the DMZ. The DMZ should be segregated from all internal network segments and resources that hold ePHI data. Further all network segments besides the DMZ should be inaccessible from external networks. Network segmentation, such as a DMZ or in conjunction with a 2 or 3-tier application configuration, is an approach for decreasing exposure and ensuring systems and data is not unnecessarily accessible by internal and/or external hosts. Information access management is a specific requirement of the HIPAA administrative safeguards – §164.308(a) (4).

- **Use of Shared Accounts:** 16% of the Healthcare Practitioner Survey respondents stated that shared accounts were in use to some degree for ePHI-relevant applications. HIPAA regulations – §164.312(a) (1) – mandate user accounts for ePHI applications be unique per individual for auditing and non-repudiation. The responses suggest that the majority of ePHI applications were using unique user accounts but not all. It was critical to review the authentication model for each ePHI application and implement user-specific accounts for any application that didn't already employ that scheme.

- **Automatic Logoff:** About 12% of the survey responses stated automatic log offs did NOT occur for all ePHI applications. HIPAA regulations – §164.312(a) (1) – require all applications that interact with ePHI to automatically log users off after a period of inactivity. According to the survey responses, a comprehensive review of all ePHI relevant applications was needed to ensure each application had this capability enabled. There was a specific comment that some applications within the hospital kept the original user logged in indefinitely, which precluded compliance.

- **Security Awareness Training:** Just under half of the survey respondents stated security awareness training was provided on a regular basis although almost as many responses indicated they were unsure if such training existed. The HIPAA regulations – §164.308(a) (5) – requires routine security awareness training and based on the responses the indication is while training exists within a portion of the organization, it is not present within all units.

A security awareness and training program needed to be established and implemented across the organization.

- **Acceptable Use Policies:** Almost 53% of the Healthcare Practitioner Survey responses indicated they were unsure if such policies existed or were signed by ePHI relevant users. HIPAA regulations – §164.308(a) (1) – mandate that acceptable use policies be created, users agree to them, and this agreement is documented prior to accessing ePHI data. Based on the responses the indication was while acceptable use policies existed within a portion of the organization, they were not pervasive through all units. Such policies need to be established that comprehensively define appropriate and inappropriate use, access, and disclosure of ePHI including sanctions for not following the policies.

- **Reporting of Unauthorized or Inappropriate ePHI Release:** Just less than half of the survey responses stated that procedures existed for reporting unauthorized or inappropriate releases of ePHI data and no responses contradicted the assertion. The other half of the responses were unsure whether such procedures existed or not. The HIPAA regulations – §164.308(a) (6) – mandate procedures be established to report and react to ePHI data being released unintentionally. While no responses indicated procedures didn't exist, the lack of understanding by the staff about such procedures in of itself created an implied deficiency. Any staff member that interacts with ePHI must understand how to identify an incident and what to do if and when they occur.

- **Physical Access Controls:** Almost 23% of the Healthcare Practitioner Survey respondents stated that physical controls were not present in all ePHI relevant locations. That indicated some areas either had no or insufficient controls for ingress/egress. The HIPAA regulations – §164.310(a) (1) – require physical access be secure and monitored. Any areas that did not have these controls in place had to be corrected.

HISCF:Phase 2

Phase 2 of the framework included an intensive technical review and assessment of the organization's IT environment. This phase measured and analyzed the actual performance of the systems and practices both against the theoretical goal presented in the HISG and the reported state of the organization provided in the assessment stage of Phase 1. It was critical for the success of this phase to identify the key IT staff within the hospital that could facilitate the exhaustive testing performed as part of the penetration and vulnerability testing. Once this staff was pinpointed, initial interviews were arranged to walk through the testing process and obtain contextual information about the environment to ensure the testing was indeed thorough but wouldn't interrupt normal business operations. It was also very important that we were able to engage directly with the manager of the EHR system to complete the EHR security and privacy assessment. Since the assessment covers a range of areas – policy, functional, and technical – it is impractical for one individual or even one group in a department to adequately respond to all questions. As such, the manager of the EHR system was able to facilitate the completion of this assessment survey. It is important to note that all systems in the organization that were hosted offsite, were considered out of scope for this phase. The technical implementation and likewise testing of those systems were implicitly regarded as meeting all compliance standards by obtaining certification from the hosting entity that their systems are compliant with the appropriate federal regulations, such as HIPAA. This is consistent with the federal government's treatment of hosted systems for audit purposes. This phase's assessments included technical interviews and inspections, penetration and vulnerability testing, and a comprehensive evaluation of security and privacy related to their EHR system. By the conclusion of Phase 2, the organization's complete IT environment had been methodically examined, tested, and documented.

Penetration and Vulnerability Testing Results

Penetration testing and vulnerability scanning by their very nature are an exhaustive, iterative process that many times requires analysis from both operational and security perspectives. One of the most common issues that lead to vulnerabilities or exploitation is merely an ignorance that a particular host is present on the network or a host is running unnecessary or unexpected services. The first step in any penetration test is to create a survey of the hosts that are present on the network and what services that are running. Many of these services are intentional and are functioning as expected. It is those hosts and related services that are unintentional that are of most significance for this initial survey. The survey portion of the security testing discovered the presence of 5,967 unique systems on the organization's production network. These hosts were running a variety of services, amongst which were SMTP, SNMP, SSL, and HTTP, which are protocols that are commonly compromised or exploited. While many of these services may serve an operational purpose, it is important to verify there are no extraneous or unexpected services operating on these ports. The partner hospital's information technology staff did examine these results and confirmed that all hosts discovered were known and the services each host was running, was intentional. An intensive battery of penetration tests and vulnerability scans were performed on the Pennsylvania hospital's production computing environment. Initially the organization's primary server subnet, subnet

A, was examined exhaustively and 98 unique hosts were discovered with 799 issues ranging from critical to low risk.

Following this assessment, the decision was made to expand the network range being tested to include other subnets that held other production and development servers as well as clients and workstations. The expanded subnets included subnets B through I. After the expanded testing was completed a total of 1,012 unique systems had been identified across the organization and 13,037 total issues of critical, high, medium, or low risk.

Based on the high number of critical and high-risk issues exposed in subnets A through I, the organization decided that a full examination of all their subnets, including those throughout the main campus that only contained workstations, would be beneficial. Following this last round of testing, 5,967 unique systems had been scanned cumulatively between all three testing exercises. In total, there were 14,448 issues found, 5,846 of which posed either a critical or high risk to the organization. The summary of the findings from all the security testing exercises are depicted in Figure 16.

Discussions and Implications

The core goal of this research was to develop potential solutions for improving accessibility, efficiency, and integrity in healthcare delivery. While this research proposes standardized approaches for evaluating and ensuring Health Insurance Portability and Accountability Act (HIPAA) compliance and for providing electronic patient access to EHR systems, these solutions needed to be tested and legitimized through actual application in a real-world environment. Chapter 4 detailed the case studies of the framework implementations with 2 national healthcare providers and the results borne out of those efforts. While the initial review of those results seems very positive, this chapter aims to delve deeper into what the results actually mean and what possible wider implications they may have for other hospitals.

The HISCF was applied with the Pennsylvania hospital, a 500-bed HIMSS 6 national hospital that admits over 400,000 patients per year. Phase 1 of the HISCF did a systematic review of the organization's policies and procedures. This phase also analyzed the hospital's technical architecture and surveyed healthcare practitioners to get a perspective on how technology was actually being used in day-to-day practice. Phase 2 of the HISCF did a thorough battery of security testing on every aspect of the Pennsylvania hospital's computing environment. Even though considered in the upper tier of hospitals in the United States with regard to information security, there were significant findings that indicated areas where Health Insurance Portability and Accountability Act (HIPAA) compliance was not being met.

Figure 16. Security issues per severity ranking

	Subnet	Unique Hosts	Unique Hosts with an...	Critical	High	Medium	Low	Totals
Data Center	A	100	98	66	234	406	93	799
Servers and Workstations (Hospital)	B	175	171	1583	2155	1611	415	5,764
	C	15	11	97	15	95	36	243
	D	205	179	24	43	1025	195	1,287
	E	205	192	0	10	1114	187	1,311
	F	209	198	15	15	1146	196	1,372
	G	183	87	126	291	603	92	1,112
	H	143	26	359	436	219	50	1,064
	I	123	50	0	54	13	18	85
	J	252	20	0	6	146	41	193
	K	40	35	38	89	253	107	487
Workstations (including Partner Practices)	L	254	18	6	33	105	30	174
	M	254	6	3	19	27	6	55
	N	254	1	0	2	7	1	10
	O	254	5	0	0	25	6	31
	P	254	3	0	0	16	3	19
	Q	254	2	0	0	12	3	15
	R	254	0	0	0	0	0	0
	S	254	0	0	0	0	0	0
	T	254	9	0	6	18	8	32
	U	254	11	0	2	51	7	60
	V	254	13	8	5	38	8	59
	W	254	0	0	0	0	0	0
	X	254	10	0	3	27	4	34
	Y	254	8	38	49	42	4	133
	Z	254	13	0	16	59	10	85
	AA	253	0	0	0	0	0	0
	BB	254	6	0	0	20	4	24
Totals		5,967	1,172	2,363	3,483	7,078	1,524	14,448

Phase 1 Inferences

While a significant number of findings were made related to the current policies, practices, and architecture of the organization' IT environment, the partner health system's level of compliance was on par with the industry averages as detailed in Figure 17.

Figure 17. Overall compliance performance

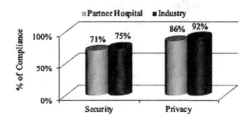

Figure 18. Compliance per functional category

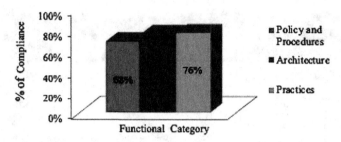

The industry averages, derived from HIMSS sponsored research (Appari, Anthony, and Johnson, 2009), indicated most organizations are closer to full compliance to privacy than security. The partner hospital mirrored this pattern with Privacy Rule compliance at 86% while the Security Rule compliance was approximately 71%. Similar to many healthcare entities, the organization was relatively close to compliance but not at the federally mandated 100% compliance as depicted in Figure 18. The functional area that required the most improvement by the organization was policy and procedures. This deficiency is fairly common throughout all industry with respect to IT and is one of the hardest areas to correct.

Changing policy and procedure requires changes to business practices and it is typically challenging for organizations to secure the leadership commitment and stakeholder buy-in to enact this type of change. Similarly, the organization had the most compliance issues with regard to the human-technology interaction element of IT compared to the four solely technical areas, as depicted in Figure 19. This was actually a good indicator for the organization that their workforce had an increasing propensity for compliance beliefs. In larger healthcare providers, over 300 beds, a

Figure 19. Compliance per technical category

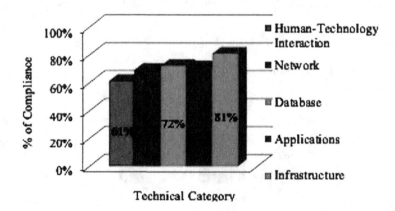

high proclivity for compliance is typically indicative of a high level of intervention by management through training, meetings, policy implementation, and enforcement. Having leadership buy-in and involvement in compliance efforts is a critical factor for an organization's compliance programs to be successful.

Even though Phase 1 yielded significant gaps in functional and technical areas that spanned the Pennsylvania hospital's computing environment, none were unsurmountable to remediate. Arguably the hardest step in compliance is simply the recognition of a requirement and corresponding discrepancy in meeting it. Ignorance through a lack of understanding and awareness is oft times the main reason for organizations be out of compliance (Kwon and Johnson, 2013). Once the issue has been identified, many times remediating the technical problem is not that difficult and can very done very quickly. This was demonstrated by the hospital's ability to respond to the majority of all of Phase 1's findings within a matter of weeks of when they were brought to their attention. Additionally, as the HISG could be leveraged for implementation level guidance, the Pennsylvania hospital's IT staff did not have to go searching for remediation solutions as they were readily available. The fact that the application of the HISCF was directly responsible the realization of the gaps in compliance and facilitated the remediation efforts suggests that the HISCF is an effective tool and was more comprehensive than the compliance program previously instituted at the Pennsylvania hospital. This claim was offered by the organization themselves after having gone through the assessment exercises laid out in Phase 1.

Phase 2 Inferences

While Phase 1 performed a passive examination of the organization's computing environment and practices, Phase 2 did an active evaluation through a series of targeted tests and inspections. Phase 2 identified 5,846 critical and high-risk issues. Undoubtedly the presence of this many elements of increased risk throughout the environment was unfortunate and disappointing to the organization. On the other hand, it was fortunate for the organization that these issues were found so they could be mitigated before the risks turned into compromises. Through analysis of the security testing results, it was discovered that many of the specific critical and high-risk vulnerabilities were found repetitively throughout the environment. Of the 5,846 critical and high-risk issues found, they are made up only 483 unique vulnerabilities.

This finding demonstrated the product of the organization not having a formal security assessment program that performed periodic testing. At a minimum it was recommended that an enterprise wide patching process and schedule be established. Additionally, it was recommended that a standardized deployment configuration for servers and workstations be developed. These fairly simple steps could mitigate many of these issues very quickly and reliably.

Figure 20. Security issues per operational host

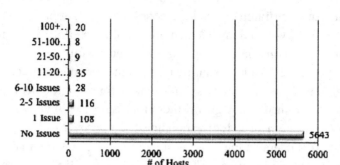

Furthermore, the routine testing would bring to light any poor implementation choices or mistakes that were made when a new system or application is brought online. The Pennsylvania hospital's technical staff was able to validate these findings and corresponding mitigation steps to resolve nearly 90% of the findings in about 1 month. Following this analysis, the results were then examined to determine how many issues each individual host had to see if there were any trends or high concentration areas of increased risk. The complete breakdown of the number of issues per host is depicted in Figure 20.

Of the 5,967 hosts present in the healthcare provider's production environment, 324 of these systems had at least 1 issue of critical or high risk. In contrast, 5,643 of the 5,967 hosts, about 95%, had no critical or high-risk issues at all. When looking only at the 1,012 machines at the hospital's main campus - subnets A through K, 725 of these machines, 72%, had no critical or high-risk issues at all. This percentage is notably similar to the organization's approximate 71% overall compliance with the HIPAA Security Rule measured in Phase 1. This similarity in results using different testing methods provides a measure of validation for the evaluation process itself as it produced comparable results between both Phases.

It is significant to note that there were 20 systems that had 100 or more unique elevated risk concerns. In fact, these 20 machines account for 4,153 of the 5,523 total critical or high-risk issues present at the main hospital campus. Which is to say just fewer than 2% of the organization's computing environment represented about 75% of its increased risk exposure. This is a common condition in most organizations as it is typical for the majority of an organization's computing environment is operating at an adequate security level and there is just a small fraction of the systems that are not.

It is concerning however that the data center, subnet A, which housed only servers (production, testing, and development), exhibited 300 critical and high-risk issues. Furthermore, only 16 of the 100 systems in this vital area of the organization and where all the storage of ePHI resided did not have at least 1 issue of critical or high risk. This means 84% of all the systems in effectively sensitive area of the environment had an elevated level of risk.

The crux of the organization's condition was a lack of true patch management program and periodic security assessment program within their computing department. As such a complete periodic security assessment program was proposed for the Pennsylvania hospital that included recommendations for all critical and high-risk issues to be measured and addressed within 30 days. It is industry-recognized that a patch management and vulnerability assessment process is a key element to mitigating risk.

This was a significant realization that came out of the testing that provided an impetus for the organization's leadership to move swiftly and decisively to correct these issues. Similarly, to Phase 1, the fact that the Pennsylvania hospital going through Phase 2 of the HISCF generated the volume and significance of results that it did, coupled with the healthcare provider's size and reputation in the industry, seems to support the claim of the effectiveness and usefulness of the HISCF.

Overall Implications of HISCF's Validity and Future Recommendations

The outcomes of applying the HISCF at the Pennsylvania hospital are very compelling and seem to suggest the framework is comprehensive and effective, at least within the scope of that hospital. Prior to applying the HISCF, the Pennsylvania hospital had failed an external audit, being cited for numerous violations related to HIPAA.

Following the HISCF implementation, the hospital was audited again, and no significant findings were reported. This suggests the HISCF was instrumental in the organization's improvement. This research recommends that further application of the HISCF is needed to strengthen the premise that it can be an effective tool for other hospitals as well. While the simple number of times the framework is applied needs to be increased, so does the variety in hospital sizes. The HISCF may not be a 'silver bullet' solution for every type of healthcare entity but the results from the case study with the Pennsylvania hospital seem indicate the potential that it could benefit other organizations around the nation and also similar healthcare organizations worldwide.

CONCLUSION

Accessibility is a pillar of healthcare delivery. However, as soon as access is afforded, it is the ethical, legal, and financial responsibility of healthcare providers to ensure the integrity of the care delivery is upheld. The Health Insurance Portability and

Accountability Act (HIPAA) and EHR systems lay the foundation for satisfying these concerns. Unfortunately, these endeavors have proved challenging to accomplish with the absence of standardized, freely available, implementation plans. Each HIPAA covered entity has been forced to approach these tasks from their localized, individual perspective and hence the figurative wheels are being reinvented again and again. Further, each one of these entities is spending vast amounts of time, resources, and money trying to determine multiple paths towards the same goals. With a lack of direction, it takes significant effort to determine what needs to be done and how to do it even before organizations can get to the point of actual implementation. As such, most healthcare organizations are expending significant and superfluous effort in the assessment and planning stages. Technology has long thrived on the adoption of standards and this research contends that the issues of accessibility, integrity, and efficiency in healthcare information technology are no exception.

There is overwhelming consensus in the healthcare industry that the spirit of HIPAA is positive and beneficial to both patients and providers. Likewise, the move from paper and film to EHR systems is clearly the natural evolution of health information storage and data exchange. It has not been so much of a struggle for most healthcare providers to find answers to the why, it has been the how that has kept these issues at the forefront of the healthcare industry for over a decade. The complexity and reach of HIPAA and the Meaningful Use programs across the entire United States has provided a seemingly endless parade of motivations for finding better methods to ensure their implementation. The guides and tools this research has produced offer promise for assisting healthcare providers with the initial implementation of these initiatives as well as better equip organizations to maintain their ongoing compliance.

REFERENCES

Acharya, S., Coats, B., Saluja, A., & Fuller, D. (2013). A Roadmap for Information Security Assessment for Meaningful Use. *Proceedings of the 2013 IEEE/ACM International Symposium on Network Analysis and Mining for Health Informatics, Biomedicine and Bioinformatics*.

Acharya, S., Coats, B., Saluja, A., & Fuller, D. (2013). Secure Electronic Health Record Exchange: Achieving the Meaningful Use Objectives. *Proceedings of the 46th Hawaii International Conference on System Sciences*, 46, pp. 253-262. 10.1109/HICSS.2013.473

Acharya, S., Coats, B., Saluja, A., & Fuller, D. (2014). From Regulations to Practice: Achieving Information Security Compliance in Healthcare. *Proceedings of the 2014 Human Computer Interaction International Conference.*

An Introductory Resource Guide for Implementing the Health Insurance Portability and Accountability Act (HIPAA) Security Rule. (2008). National Institute of Standards and Technologies. Retrieved from http://csrc.nist.gov/publications/nistpubs/800-66-Rev1/SP-800-66-Revision1.pdf

Annas, G. (2003). HIPAA Regulations - A New Era of Medical-Record Privacy? *The New England Journal of Medicine, 348*(15), 1486–1490. doi:10.1056/NEJMlim035027 PMID:12686707

Appari, A., Anthony, D. L., & Johnson, M. E. (2009). *HIPAA Compliance: An Examination of Institutional and Market Forces.* Healthcare Information Management Systems Society.

Aral, S., & Weill, P. (2007). IT assets, organizational capabilities, and firm performance: How resource allocations and organizational differences explain performance variation. *Organization Science, 18*(5), 763–780. doi:10.1287/orsc.1070.0306

Bharadwaj, S., Bharadwaj, A., & Bendoly, E. (2007). The Performance Effects of Complementarities Between Information Systems, Marketing, Manufacturing, and Supply Chain Processes. *Information Systems Research, 18*(4), 437–453. doi:10.1287/isre.1070.0148

Blumenthal, D., & Tavenner, M. (2010). The Meaningful Use Regulation for Electronic Health Records. *The New England Journal of Medicine, 363*(6), 501–504. doi:10.1056/NEJMp1006114 PMID:20647183

Breach Notification Rule. (2009). United States. Department of Health and Human Services. Office for Civil Rights. Retrieved from http://www.hhs.gov/ocr/privacy/hipaa/administrative/breachnotificationrule/index.html

45. CFR parts 160 and 164: Modifications to the HIPAA Privacy, Security, Enforcement and Breach Notification Rules Under the HITECH Act and the Genetic Information Nondiscrimination Act: Other Modifications to the HIPAA Rules: Final Rule. (2013). The United States Health and Human Services. Retrieved from http://gpo.gov/fdsys/pkg/FR-2013-01-25/pdf/2013-01073.pdf

CMS EHR Meaningful Use Overview. (2012). Center for Medicare and Medicaid Services. Retrieved from https://www.cms.gov/Regulations-and-Guidance/Legislation/EHRIncentivePrograms/Meaningful_Use.html

Coats, B., Acharya, S., Saluja, A., & Fuller, D. (2012). HIPAA Compliance: How Do We Get There? A Standardized Framework for Enabling Healthcare Information Security & Privacy. *Proceedings of the 16th Colloquium for Information Systems Security Education.*

Data and Reports. (2012). Center for Medicare and Medicaid Services. Retrieved from http://www.webcitation.org/6EMwIm36I

EMR Adoption Trends. (2014). *HIMSS Analytics.* Retrieved from http://www. himssanalytics.org/stagesGraph.asp

Enforcement Results per Year. (2010). Center for Medicare and Medicaid Services. Retrieved from http://www.hhs.gov/ocr/privacy/hipaa/enforcement/data/ historicalnumbers.html

Fichman, R., Kohli, R., & Krishnan, R. (2011). The Role of Information Systems in Healthcare: Current Research and Future Trends. *Information Systems Research, 22*(3), 419–428. doi:10.1287/isre.1110.0382

Health Reform in Action. (2010). United States White House. Retrieved from http:// www.whitehouse.gov/healthreform/healthcare-overview

Helms, M. M., Moore, R., & Ahmadi, M. (2008). Information Technology (IT) and the Healthcare Industry: A SWOT Analysis. *International Journal of Healthcare Information Systems and Informatics, 3*(1), 75–92. doi:10.4018/jhisi.2008010105

HER Incentive Programs. (2012). The Office of the National Coordinator for Health Information Technology. Retrieved from http://www.healthit.gov/providers-professionals/ehr-incentive-programs

HIPAA Administrative Simplification. (2006). United States. Department of Health and Human Services Office of Civil Rights. Retrieved from http://www.hhs.gov/ ocr/privacy/hipaa/administrative/privacyrule/adminsimpregtext.pdf

HIPAA Compliance Review Analysis and Summary of Results. (2008). Center for Medicare and Medicaid Services. Retrieved from http://www.hhs.gov/ocr/privacy/ hipaa/enforcement/cmscompliancerev08.pdf

Hirsch, R. D. (2013). Final HIPAA Omnibus Rule brings sweeping changes to health care privacy law: HIPAA privacy and security obligations extended to business associates and subcontractors. *Bloomberg Bureau of National Affairs Heath Law Reporter, 415*, 1–11.

Johnson, M. E., Goetz, E., & Pfleeger, S. L. (2009). Security through Information Risk Management. *IEEE Security and Privacy, 7*(3), 45–52. doi:10.1109/MSP.2009.77

Kayworth, T., & Whitten, D. (2010). Effective Information Security Requires a Balance of Social and Technology Factors. *MIS Quarterly Executive, 9*(3), 163–175.

Kwon, J., & Johnson, M. E. (2013). Healthcare Security Strategies for Regulatory Compliance and Data Security. *Proceedings of the 46th Hawaii International Conference on System Sciences*. 10.1109/HICSS.2013.246

Regulations and Guidance. (2004). Center for Medicare and Medicaid Services. Retrieved from https://www.cms.gov/home/regsguidance.asp

Return On Investment, H. I. P. A. A. (2005). *Blue Cross Blue Shield Association. National Committee on Vital Health Statistics*. Subcommittee on Standards and Security.

Solove, D. (2013). HIPAA Turns 10: Analyzing the Past, Present, and Future Impact. *Journal of American Health Information Management Association, 84*(4), 22–28.

Title 45 – Public Welfare, Subtitle A – Department of Health and Human Services, Part 164 – Security and Privacy. (1996). United States. National Archives and Records Administration. Retrieved from http://www.access.gpo.gov/nara/cfr/waisidx_07/45cfr164_07.html

United States Department of Commerce, National Institute of Standards and Technology. (2012). *About NSTIC*. Retrieved from http://www.nist.gov/nstic/about-nstic.html

Xia, Z., & Johnson, M. E. (2010). Access Governance: Flexibility with Escalation and Audit. *Proceedings of the 43rd Hawaii International Conference on System Sciences*.

KEY TERMS AND DEFINITIONS

HFIF: The Healthcare Federated Identity Framework aimed at positioning the healthcare providers to enable distributed electronic access to patient data.

HIPAA: The Health Insurance Portability and Accountability Act of 1996 is United States legislation that provides data privacy and security provisions for safeguarding medical information.

HISF: The Healthcare Information Security Framework aimed for organizations to plan and execute their overall HIPAA compliance projects including attestation.

HISP: The Healthcare Information Security Plan is aimed to provide comprehensive implementation level guidance for satisfying HIPAA regulations.

HISTD: The Healthcare Information Security Testing Directive and a collection of open source security testing software for organizations to assess and mitigate their systems.

Meaningful Use: The U.S. government introduced the Meaningful Use program as part of the 2009 Health Information Technology for Economic and Clinical Health (HITECH) Act, to encourage health care providers to show "meaningful use" of a certified Electronic Health Record (EHR).

Chapter 2
A Machine Learning Approach for Predicting Bank Customer Behavior in the Banking Industry

Siu Cheung Ho
The Hong Kong Polytechnic University, Hong Kong

Yuen Kwan Yau
The Hong Kong Polytechnic University, Hong Kong

Kin Chun Wong
The Hong Kong Polytechnic University, Hong Kong

Chi Kwan Yip
The Hong Kong Polytechnic University, Hong Kong

ABSTRACT

Currently, Chinese commercial banks are facing extremely tremendous pressure, including financial disintermediation, interest rate marketization, and internet finance. Meanwhile, increasing financial consumption demand of customers further intensifies the competition among commercial banks. Hence, it is very important to store, process, manage, and analyze the data to extract knowledge from the customer to predict their investment direction in future. Customer retention and fraud detection are the main information for the bank to predict customer behavior. It may involve the privacy data and sensitive data of the customer. Data security and data protection for the machine learning prediction is necessary before data collection. The research is focused on two parts: the first part is data security of machine learning and second part is machine learning prediction. The result is to prove the data security for the machine learning is important. Using different machining learning analysis tool to enhance the performance and reliability of machine learning applications, the customer behavior prediction accuracy can be enhanced.

DOI: 10.4018/978-1-5225-8100-0.ch002

INTRODUCTION

Banks was obliged to collect, analyze, and store massive amounts of data. But rather than viewing this as just a compliance exercise, machine learning and data science tools can transform this into a possibility to learn more about their clients to drive new revenue opportunities. In order to predict the customer behavior, we used machine learning algorithm by python to evaluate the customer segmentation for prediction. This section introduced the Chinese banking industry, the statement of problem in the banking industry, the project aim, objective and project scope, and project development schedule.

OVERVIEW OF CHINESE BANKING INDUSTRY

For banks globally, 2018 could be a pivotal year in accelerating the transformation into more strategically focused, technologically modern, and operationally agile institutions. They might remain dominant in a rapidly evolving ecosystem. According to Investopedia (Investopedia, 2018), the banking system in China used to be monolithic, with the People's Bank of China (PBC), which was the central bank, as the main entity authorized to conduct operations in that country. In the early 1980s, the government started opening up the banking system and allowed four state-owned specialized banks to accept deposits and conduct banking business. These five specialized banks were the Industrial & Commercial Bank of China (ICBC), China Construction Bank (CCB), Bank of China (BOC), Bank of Communications (BOC) and Agricultural Bank of China (ABC). The data security of machine learning was conducted before start the machine learning prediction.

In this work, supervised artificial neural network algorithm was implemented for classification purpose. First, challenge of Chinese banking industry was defined. Second, literature review for the machine learning approach and ANN model was evaluated. Third, data visualization and evaluation by using ANN algorithm had been analyzed for classifying the customer pattern. Fourth, the accuracy rate of customer behavior prediction was conducted. Lastly, after find tuning parameters by using XGB Classifier, the better result was awarded. Yaokai (Yaokai et al, 2018) expressed the most serious problems in recent years including problems of privacy leakage and denial of services. Early stage detection of cyber-attack was important and proposed different selection approach to test the performance of machine learning algorithm. The process had six stages. These were session splitting, feature extracting, feature ranking, cross validation, removing features gradually, and classifier.

CHALLENGE OF CHINESE BANKING INDUSTRY

Accuracy customer data prediction was essential for planning the business. After that, being armed with information about customer behaviors, interactions, and preferences, data specialists with the help of accurate machine learning models could unlock new revenue opportunities for banks by isolating and processing only this most relevant clients' information to improve business decision-making. This was a challenge to predict the customer pattern and behavior for planning the business in advance in this dynamic competition environment. Zhenyu (Zhenyu et al, 2018) stated that *"Machine learning is one of the most prevalent techniques in recent decades which has been widely applied in various fields. Among them, the applications that detect and defend potential adversarial attacks using machine learning method provide promising solutions in cybersecurity."* The application of machine learning on cybersecurity and reliability and security of machine learning system was conducted and analyzed the potential security threats against a machine learning approach in three phases of testing, training and data privacy.

PROJECT AI, OBJECTIVE AND SCOPE

This project scope included the development of Artificial Neural Network (ANN) model with 10,000 record datasets to investigate and predict which of the customers were more likely to leave the bank soon. The ANN model was developed through finding the best correlations in the dataset through data visualization technique and training a classifier which could accurately predict parameter based on the customer data. Through the use of machine learning models (Deep Learning A-Z – ANN Dataset), predict customers behaviors (e.g. exit or stay), and based on their information in the bank.

PROJECT WORK SCHEDULE AND DIVISION OF LABOUR

The project summary of the project work schedule and division of labour shown in Figure 1. It was the Gantt Chart to let the entire project member to follow the project schedule. Some project tasks started in parallel. Base on the dependencies of project task in Table 1, the activity network detail was stated in Figure 1 also.

Table 1. Project work schedule and division of labour

Task Description	Description of Goals/Benchmarks
Literature review of different Machine Learning algorithms for data analysis	Select an appropriate algorithm for the selected dataset with a sound theoretical basis
Data visualization for the selected dataset	By using visual elements like charts, graphs and plots, observe trends and identify patterns in the selected dataset.
Pre-processing of the data	Pre-processing of the data includes two main steps which were data cleaning and data transformation. Data cleaning was performed through filling in missing values and removing outliers. Data transformation referred to the normalization of the data
Build an Artificial Neural Network	Build an Artificial Neural Network in python using the pre-processed data and Keras library. The network should consisted 2-3 hidden layers using appropriate activation functions. Train the model iteratively until a reasonable good accuracy was observed (e.g. 80%).
Evaluating, improving and tuning the model	Evaluate the model performance through the test dataset. Compare the predicted result to the actual result and make appropriate adjustments to the model to improve accuracy.
Report preparation	The report was clearly identified the problem statement associated with the dataset and describe how to develop an ANN model step by step. Provide an insightful evaluation of the model and digged out some hidden patterns and relations in the dataset.
Presentation slides	A brief summarization of the problem statement, the methodologies used to approach this problem and the ANN model.

Figure 1. Gantt Chart for the work schedule and division of labor

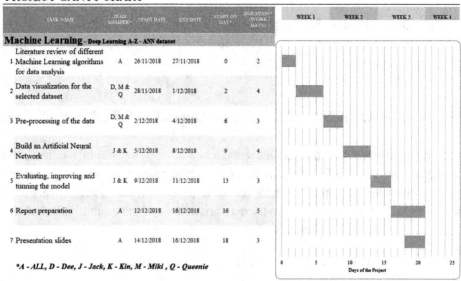

LITERATURE REVIEW

Predicting banking customer behaviors using machine learning approach to improve the prediction of customer retention in the bank was a challenging task in the worldwide banking industry. Vahid (Vahid et al., 2014) expressed that the training data and the test data was drawn from the same distribution, which was hard to be met in real world banking applications for the traditional machine learning approach. Therefore, this study evaluated the machine learning algorithms such as Artificial Neural Network (ANN) and AI platform runtime software for analyzing the workable solution to predict the customer behaviors in banking industry.

DATA SECURITY AND DATA PROTECTION

Before starting the data collection for customer behavior analysis, the data security and data protection was a critical element before using the customer data for prediction. Charles (Charles et al., 2017)) expressed the potential malicious attacks and compromised hosts may be missed in the enterprise. Machine learning was a viable approach to reduce the false positive rate and improve the productivity of Security Operation Center (SOC). Charles (Charles et al., 2017) proposed Security Information and Event Management (SIEM) System to normalize security events from different preventive technologies and flag alert to evaluate the potential malicious attacks for protecting the sensitive data in safe. Yan and J.P (Yan and J.P, 2008) introduced machine learning deals with the issue of how to build programs that improve their performance at some task through experience and suggested 6 modules to monitor the events on the computer and network such as user monitor, process monitor, network monitor, traffic monitor, file monitor and clipboard monitor. It also stated that "*Machine learning deals with the issue of how to build programs that improve their performance at some task through experience. Machine learning algorithms have proven to be of great practical value in a variety of application domains. They are particularly useful for (a) poorly understood problem domains where little knowledge exists for the humans to develop effective algorithms; (b) domains where there are large databases containing valuable implicit regularities to be discovered; or (c) domains where programs must adapt to changing conditions.*" Anna and Erhan (Anna and Erhan, 2016) combined Machine Learning (ML) and Data Mining (DM) methods for analyzing the cyber security in support of intrusion detection. The complexity of ML/DM algorithms was addressed to solve problems in six phases: 1) business understanding, 2) data understanding, 3) data preparation, 4) modeling, 5) evaluation, and 6) deployment for understanding the metrics with different information. To perform anomaly detection and misuse Intrusion Detection System (IDS) was the important datasets for training and testing the systems.

Koosha (Koosha et al., 2016) formalized a machine learning algorithm for cyber forensic and protecting the possible attack from the data collection for decision making. The key process was data acquisition, feature extraction, classification and decision making. The database information was analyzed. Attack stimulation was run to validate whether the machine learning algorithm was suitable for cyber forensic and protection. Janice and Anthony (Janice and Anthony, 2016) identified the weakest areas in Internet of Things (IoT) and proposed using machine learning within an IoT gateway to help secure the system. Carla (Carla,2017) suggested the Intelligent Cyber Security Assistant (ICSA) architecture for providing intelligent assistance to a human security specialist. The information workflow was the important role in ICSA architecture. The processes were created domain specific data model, identify features, identify and implement artificial intelligence machine learning methods, and trained implementation artificial intelligence machine learning methods. Yaokai (Yaokai et al., 2018) evaluated Command & Control (C&C) communication between compromised bots and the C&C server. It was in the preparation phase of distributed attacks and found that the detection performance was generally getting better if more features were utilized. The captioned finding was the key indicator for helping to develop the machine learning algorithms.

MACHINE LEARNING ALGORITHMS

Priyanka and Nagaraj (Priyanka and Nagaraj, 2017) stated that *"Customer churn prediction in commercial bank. They used Support Vector Machine algorithm for classification purpose. To improve the performance of the SVM model random sampling method is used and F-measure is selected for evaluating predictive power in this paper. They also developed Logistic regression model and made comparisons between developed models. The results clears that the SVM model random with sampling method works bette*r." The growing importance of analytics in banking cannot be underestimated. Machine learning algorithms and data science techniques could significantly improve bank's analytics strategy since every use case in banking was closely interrelated with analytics. As the availability and variety of information was rapidly increasing, analytics was becoming more sophisticated and accurate. Mei-Fang and Gin-Yen (Mei-Fang and Gin-Yen, 2008) investigated some key factors by using regression analysis method to predict the turnover intentions in the Taiwanese retail banking sector. Define the critical issue in advance was necessary. To better evaluate the credit risk assessment in the commercial bank, empirical analysis was critical part influence for the bank to make the decision (Guo et al., 2015).

ARTIFICAL NEURAL NETWROK (ANN)

Manging the customer relationship was a crucial problem in the telecommunications industry for analysis the customer behaviour and Customer Lifetime Value (CLV) to plan the cross selling and up-selling to the customer and building up the loyality to the company that was the improtant. Artificial neural network (ANN) was proposed by using Multi- Layer Perceptron (MLP) network with Levenberg-Marquardt algorithm to predict the CLV and the prediction result was positive (Yi et al. 2008). Customer prediction was also a extremely elements what the banking industry looking for.

The potential value of available information was astonishing: the amount of meaningful data indicating actual signals, not just noise, had grown exponentially in the past few years, while the cost and size of data processors had been decreasing. Distinguishing truly relevant data from noise contributes to effective problem solving and smarter strategic decisions. Figrue 2 was the structure of Artificial Neural Network (ANN) consists of three basic layers such as input, hidden and output layer. (Priyanka and Nagaraj 2017)

Saad (Saad et al., 2013) reviewed the cost of retaining an existing one from customer acquisition to customer retention. The results shown that "*Churn prediction has emerged as the most crucial Business Intelligence (BI) application that aims at identifying customers who are about to transfer their business to a competitor.*" ANN algorithm was used to predict the customer intention. Sanket (Sanket et al.,2018) built a multi-layered neural network for churn prediction based on customer features, support features, usage features and contextual features from data collection, pre-processing, data store, ANN classification, and then will churn or would not churn. Figure 3 was the churn neural architecture. The accuracy results were shown as 80.03%.

Figure 2. A neuron and artificial neural network

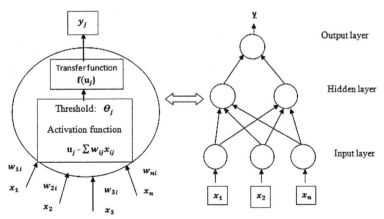

Figure 3. Churn neural architecture (Sanket et al., 2018)

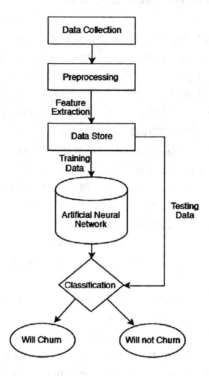

AI PLATFORM RUNTIME SOFTWARE

Yibing (Yibing et al., 2014) measured systemic risk of commercial banks in Chinese banking system by using Conditional Value-at-Risk (CoVaR) approach and extended a modified Support Vector Regression (SVR) to identify the degree of "risk externalities" for predicting systemic risk of commercials banks. Therefore, systemic prediction approach was necessary. Chang-Yun (Chang-Yun et al., 2005) stated that *"Between software architecture conception and concrete realization, semantic gap exists, the structure problem spreads to inner parts of operating system and middleware, software architecture cannot be observed and controlled by end user."* Refer to Pat Research (Pat Research, 2018), AI platform was classified as either weak AI / narrow AI, which was for a particular task or unfamiliar tasks to find out the solution. AI platform could be applied in Machine learning, automation, natural language processing and natural language understanding, and cloud infrastructure.

RESEARCH METHODOLOGY AND DATA VISUALIZATION AND EVALUATION

This chapter elaborates on the definition of data visualization, analysis the dataset of banking customer, and using ANN algorithm to evaluate the dataset through python. Data pre-processing, artificial neural network development, tuning of the artificial neural network, and data evaluation for runtime software analysis would be conducted.

DATA VISUALIZATION FOR DATASET SELECTION

The dependent variable in Table 2, the value that we are going to predict, will be the exit of the customer from the bank (binary variable 0 if the customer stays and 1 if the client exit).

The independent variables were:

- **Credit Score:** Reliability of the customer
- **Geography:** Where is the customer from
- **Gender:** Male or Female
- Age
- **Tenure:** Number of years of customer history in the company
- **Balance:** The money in the bank account
- Number of products of the customer in the bank
- **Credit Card:** If the customer has or not the CC
- **Active:** If the customer is active or not
- **Estimated Salary:** Estimation of salary based on the entries

PRE-PROCESSING DATASET

Data pre-processing was an important step in the machine learning projects particularly applicable for data-gathering methods, it often loosely controlled, resulting in out-of-range value (Data pre-processing, 2018). Through the data visualization to analysis the dataset that had not been carefully screened for such problems could produce misleading results. The fundamental analysis results were shown in Figure 4 to Figure 25. The research method was divided in 6 steps as: Step 1: Load the Dataset Information, Step 2: Analysis the Customer Segment, Step 3: Attributes with Correlation to Retention, Step 4: Attributes without correlation to Retention, Step 5: Dataset Matrix of All Attributes, and Step 6: Outliers.

Table 2. ANN dataset

Row Number	Customer Id	Surname	Credit Score	Geography	Gender	Age	Tenure	Balance	Num Of Products	HasCrCard	Is Active Member	Estimated Salary	Exited
1	15634602	Hargrave	619	France	Female	42	2	0	1	1	1	101348.88	1
2	15647311	Hill	608	Spain	Female	41	1	83807.9	1	0	1	112542.58	0
...	...	...	...	...	...	...	...	...	...	...	...	...	...

Step 1: Load the Dataset Information
Step 2: Analysis the Customer Segment

The basic information from Age Distribution, Top Ten Mode of Customer was come from age group of 31 – 40.

And also, they were come from France, Germany and Spain.

By plotting several graphs with different attributes against Retention, such as Age, Balance, No. of Product, Is Active Member, Estimated Salary, Credit Score and Has Credit Card, we found some attributes were correlated to Retention.

Figure 4. Dataset information

```
# Importing the dataset
dataset = pd.read_csv('Churn_Modelling.csv')
```

Figure 5. Age distribution

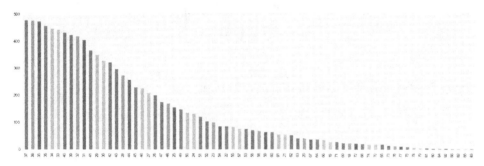

Figure 6. Distribution country

Step 3: Attributes with Correlation to Retention

1. Age vs. Retention

More than 2/3 of customer aged between 45 – 60 exited, and customer aged between the ages of 20 – 40, it was much more willing to keep their account.

2. Balance vs. Retention

Customers having Zero Balance are more willing to keep their account

3. No. of Product vs. Retention

If customer had 2 Bank's Product, great chance to retain their accounts.

4. Is Active Member vs. Retention

If customer was active member, less chance to leave the bank.

Step 4: Attributes Without Correlation to Retention

Figure 7. Attributes with correlation to retention

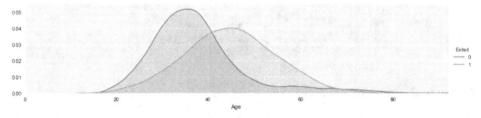

Figure 8. Customer Balance vs. Retention

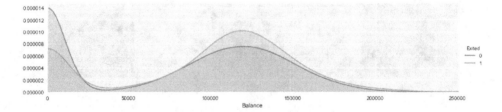

Figure 9. Number of Products vs. Retention

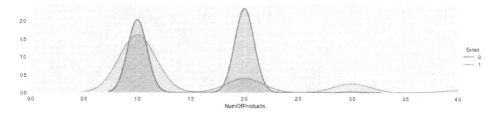

Figure 10. Active Member vs. Retention

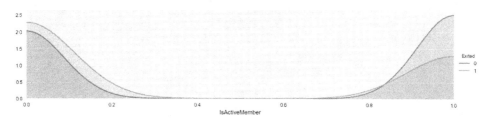

We used two curves to represent whether the customer exited or not, if two curves always overlap each other along the y-axis, it implied that attribute was not correlated to Retention.

1. Estimated Salary vs. Retention
2. Credit Score vs. Retention
3. Has Credit Card vs. Retention

Step 5: Dataset Matrix of All Attributes

From below matrix, Age vs Exited has the largest value, this means Age of customer was the most correlated attribute to Retention.

Figure 11. Estimated Salary vs. Retention

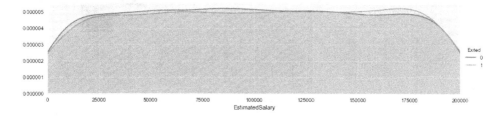

Figure 12. Credit Score vs. Retention

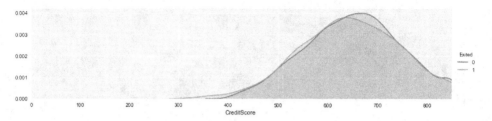

Figure 13. Has Credit Card vs Retention

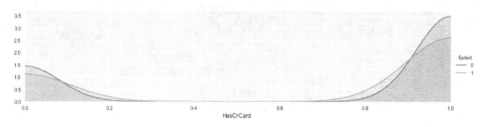

Figure 14. Dataset Matrix of All Attributes

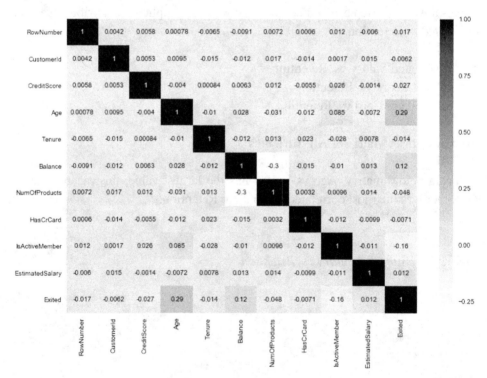

Step 6: Outliers

Usually, the accuracy of prediction was affected by some exaggerated data, thus data pre-processing was required for finding the outlier. From diagram below, two outliers were caught, one was having huge amount of Balance. Another one was having a great amount of Estimated Salary.

DATA PRE-PROCESSING

Data pre-processing was a data mining technique, which transform raw data into a specific format that was recognized by machine learning algorithm. Real-world data was often incomplete and contained many errors such as the outliers found in previous part. Therefore, data pre-processing was critical and must be done before further processing.

Scikit-learn library in python had many pre-built functionality under sklearn. Preprocessing which could be applied on our selected dataset as Figure 16, Figure 17, and Figure 18.

1. LabelEncoder function was used in order to convert 'Geography' and 'Gender' column to suitable numeric data

Figure15. Dataset Matrix of All Attributes

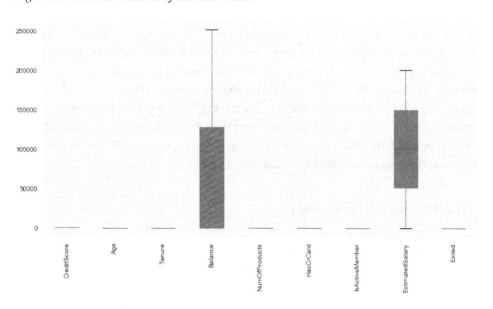

Figure 16. Convert Categorical Data to Numeric Data

```
# Encoding categorical data
X_1 = LabelEncoder()
X[:, 1] = X_1.fit_transform(X[:, 1])
X_2 = LabelEncoder()
X[:, 2] = X_2.fit_transform(X[:, 2])
onehotencoder = OneHotEncoder(categorical_features = [1])
X = onehotencoder.fit_transform(X).toarray()
X = X[:, 1:]
```

Figure 17. Data Normalisation

```
from sklearn.preprocessing import StandardScaler
StandardScaler = StandardScaler()
X_train = StandardScaler.fit_transform(X_train)
X_test = StandardScaler.transform(X_test)
```

2. StandardScaler function was used for data normalization, which scale the dataset within a specified range to prevent domination of any attribute.
3. Imblearn library was imported for the implementation of SMOTE - Synthetic Minority Over-sampling Technique to prevent the domination of one class over the other.

ARTIFICAL NEURAL NETWORK DEVELOPMENT

ANN was a computer program designed to process information, which was inspired by the human brain. A simple ANN model was trained with stochastic gradient descent and used to predict customer's behaviors in banking industry. Keras was a powerful easy-to-use Python library that contains complicated numerical computation libraries Theano and TensorFlow. It had many pre-built functionalities under Keras model and Keras layers which could be used for developing a neural network. With the help of this library, it was taken no more than a few of codes to define and train a neural network model. A fully-connected network structure with one hidden layer for this project was established and shown in below.

Figure 18. Code of Oversampling

```
#To perform over-sampling to prevent one class dominates the other
sm  = SMOTE(random_state=42)
X, y = sm.fit_sample(X, y)
```

Figure 19. A Fully-connected Network Structure

```
# Initialising the ANN
classifier = Sequential()

# Adding the input layer and the hidden layers
classifier.add(Dense(units = 6, kernel_initializer = 'uniform', activation = 'relu', input_dim = 11))
classifier.add(Dense(units = 6, kernel_initializer = 'uniform', activation = 'relu'))

# Adding the output layer
classifier.add(Dense(units = 1, kernel_initializer = 'uniform', activation = 'sigmoid'))
```

The number of input variables was 11 as specified in the **input dim** argument in the first layer because we had 11 independent variables in the dataset. Fully connected layers were defined using the 'Dense' function of Keras library. The first argument was the number of neurons in the layer, the second argument was the initialization method and the third argument was the activation function.

1. The number of neurons in the hidden layer was chosen to be ½ the size of the input layer and output layer. Therefore, the **unit** argument was set to 6.
2. A uniform weight distribution was the default initialization method in Keras, which generated a small random number between 0 and 0.05 to initialize the neural network. Therefore, **kernel_initializer** argument was set to 'uniform'.
3. 'Relu' activation function was used in the first two layers to reduce the likelihood of vanishing gradient. 'Sigmoid' activation function was used in the output layer to ensure the network output was between 0 and 1 for easy mapping.

After the fully-connected network layers were defined in the previous step, the model could be compiled using the 'compile' function as below.

Since this project was considered as a binary classification problem (e.g. exit or stay), the **loss** argument was set to 'binary_crossentropy' which could optimize the logarithmic loss for binary classification according to Keras documentations. The **optimizer** argument was chosen to be 'adam' simply for efficient computation. The model was then trained by calling the fit function using small number of iterations 100 and used a relatively small batch size of 10. These parameters were tested and modified at later stage by trial and error. After the model was fully trained, the performance of the model was evaluated using the test dataset as shown below. This gives us an overall idea of how well the model performed on new data.

Figure 20. A Fully-connected Network Layer with Compile Function

```
# Compiling the ANN
classifier.compile(optimizer = 'adam', loss = 'binary_crossentropy', metrics = ['accuracy'])

# Fitting the ANN to the Training set
classifier.fit(X_train, y_train, batch_size = 10, epochs = 100)
```

Figure 21. Diagram of Accuracy of the Model

```
#Test the model on the test dataset and compute the accuracy of model

# Predicting the Test set results
y_pred = classifier.predict(X_test)
y_pred = (y_pred > 0.5)

cm = confusion_matrix(y_test, y_pred)
ascore = accuracy_score(y_test, y_pred)
print(cm)
print("Accuracy of the model is :", ascore)
```
```
[[1412  205]
 [ 301 1268]]
Accuracy of the model is : 0.8411801632140615
```

The performance of the model was concluded to be 84.11% which was considered to be reasonably good. However, fine tuning of parameters might be able to further improve the model accuracy.

TUNING OF THE ARTIFICAL NEURAL NETWORK

Parameter optimization was a big part of machine learning because that neural network was difficult to configure with a lot of parameters which need to be set. Choosing the right parameter for the neural network was critical to the model performance. Scikit-learn library in python had many pre-built machine learning algorithms to tune the parameters of Keras deep learning models. GridSearchCV class was used as the primary parameter optimization technique for this project. A dictionary of parameters evaluation had been defined in the **parameter** argument in Figure 22. Different batch size, epochs and optimizer was analyzed in order to find the optimized parameters for our model.

The GridSearchCV function constructed and evaluated one model for each combination of parameters. Once completed, the best result observed during the optimization procedure was accessed from the **best_score_** parameter and the **best_params_** parameter described the combination that achieved the best results.

In our case, the accuracy of model after tuning of parameters was 84.45% which was not much of an increase from previous step. The parameters that achieved these results were a batch size of 25, epochs of 300 and an optimizer of 'adam'. The result proved that determining suitable training and architectural parameters of an ANN model was still remained in a difficult task. Many trials and error experiments must be done in order to find the best parameter for the model.

Figure 22. Parameter Argument

```
#Part 4 - Tuning the ANN and improving the accuracy
from keras.wrappers.scikit_learn import KerasClassifier
from sklearn.model_selection import GridSearchCV
from keras.models import Sequential
from keras.layers import Dense
def build_classifier(optimizer):
    classifier = Sequential()
    classifier.add(Dense(units = 6, kernel_initializer = 'uniform', activation = 'relu', input_dim = 11))
    classifier.add(Dense(units = 6, kernel_initializer = 'uniform', activation = 'relu'))
    classifier.add(Dense(units = 1, kernel_initializer = 'uniform', activation = 'sigmoid'))
    classifier.compile(optimizer = optimizer, loss = 'binary_crossentropy', metrics = ['accuracy'])
    return classifier
classifier = KerasClassifier(build_fn = build_classifier)
parameters = {'batch_size': [25, 32],
              'epochs': [250, 300],
              'optimizer': ['adam', 'rmsprop']}
grid_search = GridSearchCV(estimator = classifier,
                           param_grid = parameters,
                           scoring = 'accuracy',
                           cv = 10)
grid_search = grid_search.fit(X_train, y_train)
best_parameters = grid_search.best_params_
best_accuracy = grid_search.best_score_

print("Accuracy of the model is :", best_accuracy)
print("Best parameters of the model is :", best_parameters)
```

Figure 23. Best-score-parameter and Best-params-parameter

```
Epoch 298/300
12740/12740 [==============================] - 1s 75us/step - loss: 0.3342 - acc: 0.8470
Epoch 299/300
12740/12740 [==============================] - 1s 75us/step - loss: 0.3368 - acc: 0.8464
Epoch 300/300
12740/12740 [==============================] - 1s 75us/step - loss: 0.3367 - acc: 0.8488
Accuracy of the model is : 0.8445054945054945
Best parameters of the model is : {'batch_size': 25, 'epochs': 300, 'optimizer': 'adam'}
```

XGBOOST

XGBoost stands for Extreme Gradient Boosting which was developed by Tianqi Chen in 2016. It was a powerful learning algorithm based on decision tree which allowed parallel and distributed computing and an optimal usage of memory resources. XGBoost used ensemble learning method which combined several machine learning algorithms into one predictive model to achieve the best possible accuracy. Therefore, XGBoost had become a widely used machine learning algorithm among data science industry. It considered as the most useful and robust solution for applying in many real-world problems.

Python already had an in-built library for XGBoost which might be easily imported. A simple XGBoost classifier for this problem was built as shown in Figure 24.

Figure 24. XGBoost Classifier

```
import xgboost as xgb

model = xgb.XGBClassifier(max_depth = 12,random_state=7,n_estimators=100,eval_metric = 'auc' ,min_child_weight = 3
                ,colsample_bytree = 0.75, subsample= 0.8)
model.fit(X_train, y_train)
```

```
XGBClassifier(base_score=0.5, booster='gbtree', colsample_bylevel=1,
        colsample_bytree=0.75, eval_metric='auc', gamma=0,
        learning_rate=0.1, max_delta_step=0, max_depth=12,
        min_child_weight=3, missing=None, n_estimators=100, n_jobs=1,
        nthread=None, objective='binary:logistic', random_state=7,
        reg_alpha=0, reg_lambda=1, scale_pos_weight=1, seed=None,
        silent=True, subsample=0.8)
```

Parameter Setting

- **Max_depth:** This parameter represents the depth of each tree. It was found that there's no performance gain of increasing it after it was set to 12. Therefore, this value was chosen to be 12 to simplify the model and avoid overfitting.
- **Min_child_depth:** This parameter represents minimum number of samples that a node can represent in order to be split further. Since this was an imbalanced classified problem as shown in previous chapter, therefore a value of 3 was chosen for this model.
- **Subsample:** This parameter represented the percentage of rows taken to build the tree. Typical values range between 0.8-1.0, therefore a value of 0.8 was chosen for this model.
- **Colsample_bytree:** This parameter represents the percentage of columns to be randomly samples for each tree. Typical values range between 0.5-1.0, therefore a value of 0.75 was chosen for this model.
- **N_estimators:** This parameter represents the number of trees (or rounds) in an XGBoost model. The default value for this parameter was 100.
- **Eval_metric:** This parameter represents the evaluation metrics for validation of data. "auc" was chosen for this project which stands for area under the curve.

KEY FINDING AND ANALYSIS

Through the captioned finding the accuracy had been improved using XGBoost algorithm. Confusion matrix algorithm and predicting bank customer behaviors in banking industry was discussed in this study.

CONFUSIONG MARTIX

Confusion matrix (error matrix or matching matrix) was a statistical tool used to measure the accuracy and percentage of error of our machine learning algorithm. It was a 2-dimensional table comparing predicted result and actual situation. Accuracy, type I & II Errors and other measures such as precision and recall was calculated from the confusion matrix. Accuracy, type I error and type II error were calculated as a comparison of performance for the ANN and XGBoost algorithm.

Table 3 shows a typical format of a confusion matrix. Given that:

$$\text{Accuracy} = \frac{TP + TN}{TP + TN + FP + FN} \tag{1}$$

$$\text{Type I Error} = \frac{FP}{TP + FP} \tag{2}$$

$$\text{Type II Error} = \frac{FN}{FN + TN} \tag{3}$$

Table 4 showed the confusion matrix of ANN model (pre-tuning). From the above table, we knew that

Accuracy = 84.12%
Type I Error = 12.68%
Type II Error = 19.18%

Table 5 showed the confusion matrix of the XGBoost model.

Accuracy = 92.03%
Type I Error = 4.95%
Type II Error = 11.09%

Table 3. A General Format of Confusion Matrix

		Actual Class	
		Stay	Leave
Predicted Class	Stay	True Positive (TP)	False Positive (FP)
	Leave	False Negative (FN)	True Negative (TN)

Table 4. Confusion Matrix of ANN Model

		Actual Class	
		Stay	Leave
Predicted Class	Stay	1412	205
	Leave	301	1268

Table 5. Confusion Matrix of XGBoost

		Actual Class	
		Stay	Leave
Predicted Class	Stay	1537	80
	Leave	174	1395

All of the above figures were stated that XGBoost had a higher accuracy and XGBoost was more predictive than ANN model with the same data.

PREDICTING BANK CUSTOMERS BEHAVIOURS IN BANKING INDUSTRY

Predicting customer behaviors in banking industry was essential for building strong profitable customer relationship through managing the customer impression and trust. ANN model was used to predict the customer behavior from the historical data to analysis the customer pattern through this single factor basic for prediction.

SINGLE-FACTOR BASIC

On a single-factor basis, it was observed from the characteristics described in part 3, these were Age, Balance and Earnings and Salary. The attributes were the most influential in deciding whether a customer would stay in the bank, supporting by plots and correlation table. However, it was observed that the correlation was not significant as the highest absolute correlation to 0.29. A single factor for prediction was evaluated to be prone to error. ANN model was hence proved a more appropriate solution.

ANN MODEL

ANN model was used to configure the application of a set of input and train by feeding the customer patterns and analysis the weights and segments for classification and prediction as well as clustering methods.

Figure 25 and Figure 26 showed the tuning parameters and results of the ANN model respectively. Using the parameters, the optimized model was concluded to 84.45% of accuracy:

- Batch size equals to 25. It was the minimum batch size of our option and suggested that as the records present were relatively small, a smaller batch size that might be led to a faster converge rate.
- There were 300 epochs. It was noted that more epochs could help drawing higher accuracy, while it might face a plateau after reaching a certain number of epochs. In this project, 300 epochs was observed and the performance was better on the same basis and that the accuracy rate was increased from 250 epochs to 300 epochs.
- Adam was the optimizer. The effect of Adam and RMSprop were usually similar. While Adam was an optimizer built on the basis of RMSprop with the addition of bias-correction and momentum. In our algorithm, Adam works was better in combination with another parameter.

Figure 25. Parameters Set for Tuning

```
#Part 4 - Tuning the ANN and improving the accuracy
from keras.wrappers.scikit_learn import KerasClassifier
from sklearn.model_selection import GridSearchCV
from keras.models import Sequential
from keras.layers import Dense
def build_classifier(optimizer):
    classifier = Sequential()
    classifier.add(Dense(units = 6, kernel_initializer = 'uniform', activation = 'relu', input_dim = 11))
    classifier.add(Dense(units = 6, kernel_initializer = 'uniform', activation = 'relu'))
    classifier.add(Dense(units = 1, kernel_initializer = 'uniform', activation = 'sigmoid'))
    classifier.compile(optimizer = optimizer, loss = 'binary_crossentropy', metrics = ['accuracy'])
    return classifier
classifier = KerasClassifier(build_fn = build_classifier)
parameters = {'batch_size': [25, 32],
              'epochs': [250, 300],
              'optimizer': ['adam', 'rmsprop']}
grid_search = GridSearchCV(estimator = classifier,
                           param_grid = parameters,
                           scoring = 'accuracy',
                           cv = 10)
grid_search = grid_search.fit(X_train, y_train)
best_parameters = grid_search.best_params_
best_accuracy = grid_search.best_score_

print("Accuracy of the model is :", best_accuracy)
print("Best parameters of the model is :", best_parameters)
```

Figure 26. Result of Tuning

```
Accuracy of the model is : 0.8445054945054945
Best parameters of the model is : {'batch_size': 25, 'epochs': 300, 'optimizer': 'adam'}
```

CONCLUSION AND RECOMMENDATION

The privacy data and sensitive data of the customer were important and must need to be secured. Data security and data protection for the machine learning prediction was necessary before data collection and analysis. Real-time analytics helped to understand the problem that held back the business, while predictive analytics aid in selecting the right technique to solve it. Significantly better results were achieved by integrating analytics into the bank workflow to avoid potential problems in advance. In this study, ANN model was used to try to predict and analyze if a customer would maintain its account in the bank, determining by the customers' information such as age and balance. The accuracy of 84.12% accuracy had been achieved before tuning. ANN model was using unsupervised learning method to the data to analyze the dataset to cluster the customer pattern of normal and abnormal observations. This study reviewed the pros and cons of machine learning model. The parameters for prediction accuracy improvement were analyzed.

PROS AND CONS TO BE CONSIDERED

Pros and Cons evaluation was helping the decision-making for the company. It was regarded as a problem-solving activity terminated by a solution deemed to be optimal, or at least satisfactory. Table 6 was a summary to be considered (Jack V Tu, 1996).

IMPROVEMENT AND FIND TUNING PARAMETERS

XGBoost was a part of the Distributed (Deep) Machine Learning Community (DMLC) group. It became well known in the ML competition circles. This brought the library to more developers and became popular among the Kaggle community where it had been used for a large number of competitions (XGBoost, 2018). To enhance the accuracy of the model, XGBoost was used to find tuning the parameters for evaluating the dataset. Finally, the results of accuracy were increased to 92.03% and shown in Figure 27.

Table 6. Pros and Cons Evaluation

Pros	Cons
❖ ANN can model non-linear data	❖ ANN model was computation intensive.
❖ Non-linear data was suggested to be significantly more challenging as it was often more difficult to seek for a statistical and mathematical explanation.	❖ It set a higher demand for the hardware requirements
❖ ANN offered a new solution for non-linear modelling on top of current solution such as non-linear regression model, which based on interpolations.	❖ ANN model might be more expensive in computation time and hardware requirement than other models.
	❖ ANN model required a relatively big data set
	❖ In order to increase the accuracy, it was required to have a minimum amount gradient updates, ANN generally tilts to be more realistic for big data.

Figure 27. Results of XGBClassifier for Improving Parameters

```
In [21]:  import xgboost as xgb

          model = xgb.XGBClassifier(max_depth = 12,random_state=7,n_estimators=100,eval_metric = 'auc' ,min_child_weight = 3
                                    ,colsample_bytree = 0.75, subsample= 0.8)
          model.fit(X_train, y_train)

Out[21]:  XGBClassifier(base_score=0.5, booster='gbtree', colsample_bylevel=1,
                colsample_bytree=0.75, eval_metric='auc', gamma=0,
                learning_rate=0.1, max_delta_step=0, max_depth=12,
                min_child_weight=3, missing=None, n_estimators=100, n_jobs=1,
                nthread=None, objective='binary:logistic', random_state=7,
                reg_alpha=0, reg_lambda=1, scale_pos_weight=1, seed=None,
                silent=True, subsample=0.8)

In [22]:  y_pred = model.predict(X_test)
          cm = confusion_matrix(y_test, y_pred)
          print(cm)
          print("Accuracy of the model is :", accuracy_score(y_test, y_pred))

          [[1537   80]
           [ 174 1395]]
          Accuracy of the model is :  0.9202762084118016
```

PARAMETERS

Adam and RMSprop was common optimizer for ANN model. Stochastic Gradient Descent (SGD), Momentum and AdaGrad were some other optimizers that added to the parameters for tuning if computation power was adequate. A set of higher number of epochs was tested. In this case study, it was observed that more cases of epochs larger than 300 can be tested. After the model was fit and used to predict the test dataset, it was found that achievement was a much higher accuracy rate of 92%. It was also observed that this XGBoost model required much less computational power to run. It was well known that XGBoost algorithm was currently the most useful algorithm in data science.

REFERENCES

Behbood, V., Lu, J., & Zhang, G. (2014). Fuzzy Refinement Domain Adaptation for Long Term Prediction in Banking Ecosystem. *IEEE Transactions on Industrial Informatics*, 10(2), 1637 – 1646. 10.1109/TII.2012.2232935

Buczak, A. L., & Guven, E. (2016). A Survey of Data Mining and Machine Learning Methods for Cyber Security Intrusion Detection. *IEEE Communications Surveys and Tutorials*, *18*(2), 1153–1176. doi:10.1109/COMST.2015.2494502

Ca˜nedo, J., & Skjellum, A. (2016). Using machine learning to secure IoT systems. *IEEE Conference Publications*, 219 – 222. 10.1109/PST.2016.7906930

Chen, Shi, Lee, Li, & Liu. (2014). The Customer Lifetime Value Prediction in Mobile Telecommunications. *IEEE Conference Publications*, 565 – 569.

Chen, M.-F., & Lien, G.-Y. (2008). The Mediating Role of Job Stress in Predicting Retail Banking Employees' Turnover Intentions in Taiwan. *IEED Conference Publications*, 393 - 398.

Data Pre-Processing. (2018). In *Wikipedia*. Retrieved from https://en.wikipedia.org/wiki/Data_pre-processing

Deep Learning A-Z - ANN dataset. (n.d.). Retrieved from https://www.kaggle.com/filippoo/deep-learning-az-ann)

Feng, Y., Akiyama, H., Lu, L., & Sakurai, K. (2018). Feature Selection For Machine Learning-Based Early Detection of Distributed Cyber Attacks. *IEEE Conference Publications*, 173 – 180. 10.1109/DASC/PiCom/DataCom/CyberSciTec.2018.00040

Guan, Z., Bian, L., Shang, T., & Liu, J. (2018). When Machine Learning meets Security Issues: A survey. *IEEE Conference Publications*, 158 - 165. 10.1109/IISR.2018.8535799

Investopedia. (2018). *Introduction to the Chinese Banking*. Retrieved from https://www.investopedia.com/articles/economics/11/chinese-banking-system.asp

Li, C.-Y., Jiang, L., Liang, A.-N., & Liao, L.-J. (2005). A User-Centric Machine Learning Framework for Cyber Security Operations Center. *IEEE Conference Publications*, 173-175.

Luo, Y., & Tsai, J. J. P. (2008). A Framework for Extrusion Detection Using Machine Learning. *IEEE Conference Publications*, 83 – 88.

Pat Research. (2018). *Artificial Intelligence Platforms*. Retrieved from https://www.predictiveanalyticstoday.com/artificial-intelligence-platforms/

Paul, P. S., & Dharwadkar, N. V. (2017). Analysis of Banking Data Using Machine Learning. *IEEE Conference Publications*, 876 - 881.

Qureshi, S. A., Rehman, A. S., Qamar, A. M., Kamal, A., & Rehman, A. (2013). Customer Churn Prediction Modelling Based on Behavioural Patterns Analysis using Deep Learning. *IEEE Conference Publications*, 1 - 6.

Sadeghi, K., Banerjee, A., Sohankar, J., & Gupta, S. K. S. (2016). Toward Parametric Security Analysis of Machine Learning based Cyber Forensic Biometric Systems. *IEEE International Conference on Machine Learning and Applications*, 626 – 631. 10.1109/ICMLA.2016.0110

Sayan, C. M. (2017). An Intelligent Security Assistant for Cyber Security Operations. *IEEE Conference Publications*, 375 – 376. 10.1109/FAS-W.2017.179

Tu. (1996). Advantages and disadvantages of using artificial neural networks versus logistic regression for predicting medical outcomes. *Journal of Clinical Epidemiology*. Retrieved from https://www.sciencedirect.com/science/article/pii/S0895435696000029

Wei, Yingjie, & Mu. (2015). Commercial Bank Credit Risk Evaluation Method based on Decision Tree Algorithm. *IEEE Conference Publications*, 285 -288.

XGBoost. (2018). In *Wikipedia*. Retrieved from https://en.wikipedia.org/wiki/XGBoost

Chapter 3
Advanced-Level Security in Network and Real-Time Applications Using Machine Learning Approaches

Mamata Rath
ⓘD https://orcid.org/0000-0002-2277-1012
Birla Global University, India

Sushruta Mishra
KIIT University (Deemed), India

ABSTRACT

Machine learning is a field that is developed out of artificial intelligence (AI). Applying AI, we needed to manufacture better and keen machines. Be that as it may, aside from a couple of simple errands, for example, finding the briefest way between two points, it isn't to program more mind boggling and continually developing difficulties. There was an acknowledgment that the best way to have the capacity to accomplish this undertaking was to give machines a chance to gain from itself. This sounds like a youngster learning from itself. So, machine learning was produced as another capacity for computers. Also, machine learning is available in such huge numbers of sections of technology that we don't understand it while utilizing it. This chapter explores advanced-level security in network and real-time applications using machine learning.

DOI: 10.4018/978-1-5225-8100-0.ch003

INTRODUCTION

Machine Learning is a recent development in the area of science and technology which is based on the foundation of Artificial Intelligence(AI). By applying AI, we needed to manufacture better and improved machines. Be that as it may, aside from couple of simple errands, for example, finding the briefest way between two points, it isn't to program more mind boggling and continually developing difficulties. There was an acknowledgment that the best way to have the capacity to accomplish this undertaking was to give machine a chance to gain from itself. This sounds like a technically similar learning from its self. So machine learning was produced as another capacity for computers. Also, now machine learning is available in such huge numbers of sections of technology, that we don't understand it while utilizing it.

Machine learning (ML) is also concerned about the structure and advancement of network security and strategies that enables systems to learn and train. The significant focal point of machine learning explore is to extricate data from information consequently, by computational and measurable techniques. It is subsequently firmly identified with information mining and insights. The intensity of neural networks originates from their portrayal ability. From one viewpoint, feed forward networks are demonstrated to offer the ability of general capacity guess. Then again, intermittent networks utilizing the sigmoidal initiation work are Turing proportionate and recreates a general Turing machine; Thus, repetitive networks can figure whatever work any advanced computer can register.

Discovering designs in information on planet earth is conceivable just for human minds. The information being extremely gigantic, the time taken to register is expanded, and this is the place Machine Learning comes enthusiastically, to assist individuals with vast information in least time. On the off chance that enormous information and distributed computing are gaining significance for their commitments, machine learning as technology breaks down those huge lumps of information, facilitating the errand of information researchers in a computerized procedure and gaining square with significance and acknowledgment. The methods we use for information digging have been around for a long time, however they were not viable as they didn't have the focused capacity to run the calculations. In the event that we run profound learning with access to better information, the yield we get will prompt emotional leaps forward which is machine learning.

This chapter has been organised as follows. Section 1 depicts the Introduction part. Section 2 illustrates Security in Network and Solution in Machine Learning, section 3 focuses on Cyber attacks in IoT and Cloud Based machine learning, section 4 highlights Security and Vulnerability in Wireless Network due to various attack, section 5 details about Assortment of Machine Learning Practice for Security & Analysis, section 6 describes Risk Assessment in IoT Network and at last section 7 concludes the chapter.

SECURITY IN NETWORK AND SOLUTION IN MACHINE LEARNING

Malware investigation and categorization Systems utilize static and dynamic methods, related to machine learning calculations, to computerize the assignment of ID and grouping of malevolent codes. The two procedures have shortcomings that permit the utilization of analysis avoidance systems, hampering the ID of malwares. R. J. Mangialardo et.al,(2015) propose the unification of static and dynamic analysis, as a strategy for gathering information from malware that reductions the possibility of achievement for such avoidance strategies. From the information gathered in the analysis stage, we utilize the C5.0 and Random Forest machine learning calculations, actualized inside the FAMA structure, to play out the distinguishing proof and order of malwares into two classes and various classifications. The examinations and results demonstrated that the exactness of the bound together analysis accomplished a precision of 95.75% for the double arrangement issue and an exactness estimation of 93.02% for the different order issue. In all examinations, the brought together analysis created preferred outcomes over those acquired by static and dynamic breaks down detached.

Safeguard for Mobile Communication

A novel way to deal with ensuring cell phones has been arranged (N. Islam et.al, 2017) from malware that may release private data or adventure vulnerabilities. The methodology, which can likewise shield gadgets from interfacing with pernicious passageways, utilizes learning strategies to statically investigate applications, examine the conduct of applications at runtime, and screen the manner in which gadgets connect with Wi-Fi passageways.

Intrusion Detection System Using Machine Learning Approach

Intrusion detection is an essential section of security system such as versatile security apparatuses, intrusion detection frameworks, intrusion counteractive action frameworks, and firewalls. Different intrusion detection strategies are utilized, yet their execution is an issue. Intrusion detection execution relies upon precision, which needs to enhance to diminish false alerts and to expand the detection rate.

To determine worries on execution, multilayer perceptron, bolster vector machine (SVM), and different procedures have been utilized in recent work. Such methods show impediments and are not productive for use in substantial informational collections, for example, framework and system information. The intrusion detection framework is utilized in dissecting gigantic activity information; along these lines, a productive arrangement system is important to beat the issue. This issue is considered by I.Ahmed et.al (2018) utilizing understood machine learning procedures, in particular, SVM, irregular woods, and outrageous learning machine (ELM) are connected. These systems are notable due to their capacity in grouping. The NSL-learning disclosure and information mining informational collection is utilized, or, in other words benchmark in the assessment of intrusion detection components. The outcomes demonstrate that ELM outflanks different methodologies.

Detection of Cyber Attacks

Attack detection issues in the radiant framework are acted like factual learning issues for various attack situations in which the estimations are seen in clump or online settings. In this methodology, machine learning calculations are utilized (M.Ozay et.al, 2016) to characterize estimations as being either secure or attacked. An attack detection system is given to abuse any accessible earlier information about the framework and surmount imperatives emerging from the meagre structure of the issue in the proposed methodology. Surely understood clump and web based learning calculations (directed and semisupervised) are utilized with choice and highlight level combination to model the attack detection issue. The connections among measurable and geometric properties of attack vectors utilized in the attack situations and learning calculations are broke down to recognize imperceptible attacks utilizing factual learning techniques. The proposed calculations by (M.Ozay et.al, 2016)are analyzed on different IEEE test frameworks. Trial examinations demonstrate that machine learning calculations can identify attacks with exhibitions higher than attack detection calculations that utilize state vector estimation strategies in the proposed attack detection structure.

CYBER ATTACKS IN IOT AND MACHINE LEARNING STRATEGY

The development and advancement of cyber-attacks require strong and developing cyber security plans. As a developing innovation, the Internet of Things (IoT) acquires cyber-attacks and dangers from the IT condition in spite of the presence of a layered guarded security instrument. The augmentation of the computerized world to the physical condition of IoT brings inconspicuous attacks that require a novel lightweight and conveyed attack detection system because of their engineering and asset limitations. Compositionally, Fog computing based mobile stations can be utilized to offload security capacities from IoT and the cloud to moderate the asset restriction issues of IoT and versatility bottlenecks of the cloud. Traditional machine learning calculations have been widely utilized for intrusion detection, despite the fact that versatility, highlight designing endeavors, and precision have prevented their infiltration into the security advertise. These inadequacies could be alleviated utilizing the profound learning approach as it has been fruitful in huge information fields. Aside from disposing of the need to create includes physically, profound learning is strong against transforming attacks with high detection exactness. A. Diro et.al, (2018) proposed a LSTM arrange for circulated cyber-attack detection in mist to-things communication. Critical attacks have been investigated and dangers focusing on IoT gadgets were distinguished particularly attacks abusing vulnerabilities of remote correspondences. The directed investigations on two situations show the adequacy and productivity of more profound models over conventional machine learning models.

Non-Reliable Data Source Identification Using Machine Learning Algorithm

Recent advances in machine learning have prompted imaginative applications and administrations that utilization computational structures to reason about complex marvel. In the course of recent years, the security and machine-learning networks have created novel methods for developing ill-disposed examples - malicious data sources made to deceive and in this manner degenerate the trustworthiness of frameworks based on computationally learned models. The hidden reasons for antagonistic examples and the future countermeasures has been broke down (P.McDaniel et.al, 2016) that may relieve them.

Deep Learning and Machine Learning for Interruption in Network

With the improvement of the Internet, cyber-attacks are changing quickly and the cyber security circumstance isn't idealistic. Overview report by Y.Xin et.al (2018) clarifies the key writing studies on machine learning (ML) and deep learning (DL) techniques for system enquiry of interruption identification and gives a concise instructional exercise portrayal of every ML/DL strategy. Distinctive security approaches were ordered and outlined dependent on their transient or warm connections. Since information are so essential in ML/DL strategies, it portrays a portion of the generally utilized system datasets utilized in ML/DL, talk about the difficulties of utilizing ML/DL for cyber security and give recommendations to inquire about bearings.

Security Guarded Procedures Using Machine Learning

Machine learning is a standout amongst the most overall procedures in software engineering, and it has been generally connected in picture preparing, regular dialect handling, design acknowledgment, cyber security, and different fields. Notwithstanding fruitful utilizations of machine learning calculations in numerous situations, e.g., facial acknowledgment, malware location, programmed driving, and interruption discovery, these calculations and comparing preparing information are helpless against an assortment of security dangers, initiating a critical execution diminish. Consequently, it is indispensable to call for further consideration with respect to security dangers and comparing guarded procedures of machine learning, which persuades a complete review (Q.Liu et.al, 2018). Up to this point, specialists from the scholarly community and industry have discovered numerous security dangers against an assortment of learning calculations, including credulous Bayes, strategic relapse, choice tree, bolster vector machine (SVM), rule part examination, bunching, and winning profound neural systems.

There are many implementations of machine learning approach that utilizes supervisory learning. In supervised learning, the framework attempts to gain from the past precedents that are given. (Then again, in unsupervised learning, the framework endeavors to discover the examples straightforwardly from the model given.) Speaking scientifically, regulated learning is the place you have both info factors (x) and yield variables(Y) and can utilize a calculation to get the mapping capacity from the contribution to the yield. Regulated learning issues can be additionally partitioned into two sections, in particular characterization, and relapse.

A classification issue is the dilemma at which the yield variable is a classification or a gathering, for example, "dark" or "white" or "spam" and "no spam". Regression: A regression issue is the point at which the yield variable is a genuine esteem, for example, "Rupees" or "stature." Unsupervised Learning - In unsupervised learning, the calculations are left to themselves to find fascinating structures in the information. Scientifically, unsupervised learning is the point at which you just have input information (X) and no relating yield factors. This is called unsupervised learning in light of the fact that not at all like directed learning above, there are no given right answers and the machine itself finds the appropriate responses. Unsupervised learning issues can be additionally separated into association and grouping issues. Association: An association rule learning issue is the place you need to find decides that depict substantial parts of your information, for example, "individuals that purchase X additionally will in general purchase Y". A clustering issue is the place you need to find the innate groupings in the information, for example, gathering clients by buying conduct.

- **Reinforcement Learning:** A computer program will communicate with a dynamic situation in which it must play out a specific objective, (for example, playing a diversion with a rival or driving a vehicle). The program is given criticism regarding prizes and disciplines as it explores its concern space. Utilizing this algorithm, the machine is prepared to settle on explicit choices. It works along these lines: the machine is presented to a situation where it consistently prepares itself utilizing experimentation technique.

Machine Learning supposition is a field that meets factual, probabilistic, computer science and algorithmic angles emerging from learning drearily from information which can be utilized to assemble savvy applications. The preeminent inquiry when attempting to comprehend a field, for example, Machine Learning is the measure of maths important and the unpredictability of maths required to comprehend these frameworks. The response to this inquiry is multidimensional and relies upon the dimension and enthusiasm of the person. Here is the base dimension of science that is required for Machine Learning Engineers/Data Scientists.Machine learning approaches are basically used in mathematical fields such as linera algebra including matrix operations, projections, factorisation, symmetric matrix and orthogonalisation. In Probability and statistics it includes rules and axioms, bayes'theorem, random variables, variance, expectation, conditional and joint distributions. In calculus, differential and integral calculus and partial derivatives are implemented in machine learning approachs.Further Design of Algorithm and complex optimisations includes binary tree, hashing, heap and stack operations.

Figure 1. Reinforcement in Machine Learning

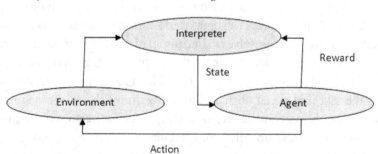

Methods in Neural Networks

It is obvious the learning speediness of feed forward neural networks is all in all far slower than required and it has been a noteworthy bottleneck in their applications for past decades. Two key purposes for might be: (1) the moderate gradient based learning calculations are broadly used to prepare neural networks, and (2) every one of the parameters of the networks are tuned ordinarily by utilizing such learning calculations. FFNN (Feed forward Neural Networks) are most widely utilized in numerous fields because of their capability such as (1) to estimated complex nonlinear mappings straightforwardly from the information tests; and (2) to give models to a substantial class of characteristic and counterfeit wonders that are hard to deal with utilizing traditional parametric methods. Then again, there need quicker learning calculations for neural networks. The conventional learning calculations are more often than not far slower than required. It isn't astonishing to see that it might take a few hours, a few days, and significantly more opportunity to prepare neural networks by utilizing customary techniques.

From a numerical perspective, look into on the estimation capacities of feedforward neural networks has concentrated on two angles: all inclusive guess on conservative information sets and estimation in a limited arrangement of preparing tests. Numerous analysts have investigated the all inclusive guess capacities of standard multilayer Feed Forward neural networks. It was demonstrated that in the event that the enactment work is nonstop, limited and nonconstant, ceaseless mappings can be approximated in measure by neural networks over minimized information sets. It was again demonstrated that feedforward networks with a nonpolynomial enactment capacity can inexact (in measure) constant capacities. In genuine applications, the neural networks are prepared in limited preparing set. For capacity estimation in a limited preparing set, a novel approach shows that a Solitary concealed Layer Feed forward Neural network (SLFN) with at most N shrouded nodes and with any nonlinear actuation capacity can precisely learn N unmistakable perceptions. It ought to be noticed that the information weights (connecting the information layer to the main concealed layer) and shrouded layer predispositions should be balanced in all these past hypothetical research functions and in addition in all handy learning calculations of feedforward neural networks.

Normally, every one of the parameters of the feedforward networks should be tuned and in this manner there exists the reliance between various layers of parameters (weights and predispositions). For past decades, inclination drop based techniques have principally been utilized in different learning calculation of feed forward neural networks. Be that as it may, unmistakably slope plunge based learning techniques are commonly ease back because of inappropriate learning steps or may effectively combine to nearby minima. Also, numerous iterative learning steps might be required by such learning calculations with the end goal to acquire better learning execution.

Malware Detection Using Machine Learning

In spite of the huge enhancement of digital security instruments and their ceaseless advancement, malware are still among the best dangers in the internet. Malware examination applies methods from a few distinct fields, for example, program investigation and network examination, for the investigation of pernicious examples to build up a more profound comprehension on a few viewpoints, including their conduct and how they advance after some time. Inside the constant weapons contest between malware designers and experts, each development in security technology is normally speedily pursued by a relating avoidance. Some portion of the viability of novel cautious measures relies upon what properties they use on. For instance, a recognition rule dependent on the MD5 hash of a known malware can be effortlessly evaded by applying standard systems like jumbling, or further developed methodologies, for example, polymorphism or changeability. For a complete survey of these procedures.. These techniques change the double of the malware, and hence its hash, yet leave its conduct unmodified. On the opposite side, creating identification decides that catch the semantics of a noxious example is considerably more hard to evade, in light of the fact that malware engineers ought to apply more mind boggling changes(Rath et.al, 2018). A noteworthy objective of malware investigation is to catch extra properties to be utilized to enhance safety efforts and make avoidance as hard as would be prudent. Machine learning is a characteristic decision to help such a procedure of information extraction. In fact, numerous works in writing have taken this bearing, with an assortment of methodologies, goals and results.

SECURITY AND VULNERABILITY IN WIRELESS NETWORK DUE TO VARIOUS ATTACK

In wireless network, associated devices such as laptops, PCs, cellular phones, appliances with communication capability are linked together to create a network. MANET is a self-arranging system of versatile switches related hosts associated by remote connections. The routers (mobile gadgets) move haphazardly and compose themselves self-assertively; along these lines, the systems remote topology may change quickly and capriciously (Rath et.al, 2018) . In MANETs each node acts as router and because of dynamic changing topology the accessibility of hubs is not generally ensured. It likewise does not ensure that the way between any two hubs would be free of pernicious hubs. The remote connection between hubs is exceptionally vulnerable to connection assaults such as passive eavesdropping, active interfering, etc.Stringent asset limitations in MANET may likewise influence the nature of security when excessive computations are required to perform some encryption(Rath et.al, 2018) . These vulnerabilities and characteristics make a case to build a security solution which provides security services like authentication, confidentiality, integrity, non-repudiation and availability. In order to achieve this goal we need a mechanism that provides security in each layer of the protocol.Various attacks on Routing Protocols in wireless networks are as follows.

1. Black Hole Attack
2. Wormhole Attack
3. Rushing Attack
4. **Passive Attacks:** The attacker just spies around the network without distracting the network operation. This attack compromises the privacy of the data and says which nodes are working in immoral way.
5. **Active Attacks:** It is a type of attack in which the attacker disturbs the normal operation of the network by fabricating messages, dropping or changing packets, by repeating or channelling them to other part of the network (Rath et.al, 2018). Basically, the content of the message is changed. It is of two types:
6. **External Attacks:** Here the attacker causes network jamming and this is done by the propagation of fake routing information. The attack disrupts the nodes to gain services.
7. **Internal Attacks:** Here the attacker wants to gain access to network and wants to get involved in network activities. Attacker does this by some malicious imitation to get access to the network as a new node or by directly through a current node and using it as a basis to conduct the attack.

Black Hole Attack

Worm hole attack-Malicious nodes eavesdrops the packets, tunnel them to another location in the network and retransmit them at the other end. Fig.2 black hole attack in mobile wireless network and Fig.3. shows Worm hole attack in wireless network.

Rushing Attack

Forward ROUTE Requests more quickly than legitimate nodes can do so, increase the probability that routes that include the attacker will be discovered, attack against all currently proposed on-demand ad hoc network routing protocols.

Figure 2. Black hole attack in mobile wireless network

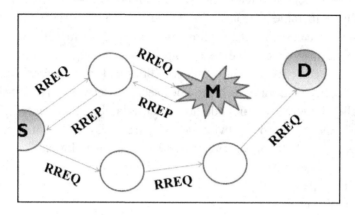

Figure 3. Worm hole attack in wireless network

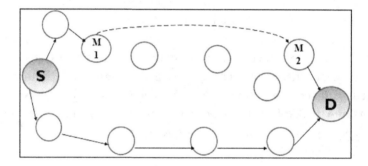

Collaborative Attacks

Collaborative attacks (CA) occur when more than one attacker synchronize their actions to disturb a target network. Different Models of Collaborative Attack

- Collaborative Black hole attack
- Collaborative Black hole and Wormhole attack
- Collaborative Black hole and Rushing Attack

Fig.4., Fig 5 and Fig.6. show different Collaborative black hole attacks. Collaborative black hole and worm hole attack- Current Proposed Solutions to handle collaborative black hole attack are (a). Collacorative Monitoring: Collaborative security architecture for black hole attack prevention (b).Recursive Validation - Prevention of Cooperative Black Hole Attack in wireless Networks.

Figure 4. Collaborative black hole attack type 1

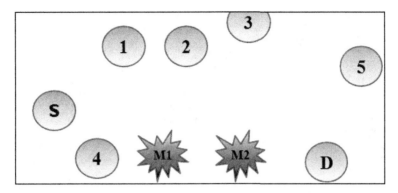

Figure 5. Collaborative black hole attack type 2

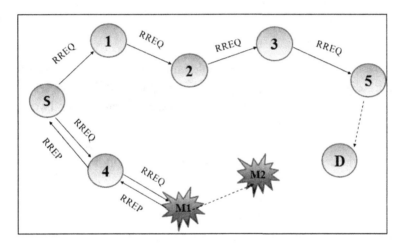

Figure 6. Collaborative black hole attack type 3

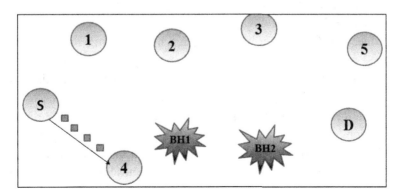

Figure 7. Collaborative black hole and worm hole attack

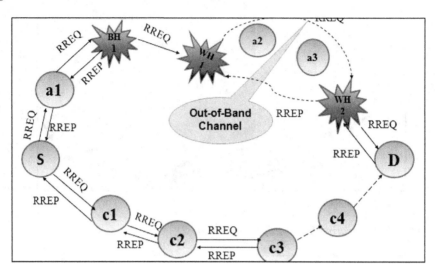

Fig.6 presents Collaborative black hole and worm hole attack. Monitoring is done during data transmission and loss of data packets take place. The current solutions does not specify if and how the lost data is re-transmitted. Two important overhead in Monitoring even if no attack is present, and in isolating the malicious nodes recursively. The solution is to get a count of the packets received from the destination. If the count is less than a threshold then monitor.

ASSORTMENT OF MACHINE LEARNING PRACTICE FOR SECURITY AND ANALYSIS

The most widely recognized goal with regards to malware analysis is distinguishing whether a given example is malevolent. This goal is additionally the most essential since knowing ahead of time that an example is perilous permits to square it before it winds up unsafe. Without a doubt, the greater part of surveyed works has this as principle objective. Contingent upon what machine learning strategy is utilized, the produced yield can be furnished with a certainty esteem that can be utilized by examiners to comprehend if an example needs further examination. Another significant goal is spotting similitude among malware, for instance to see how novel examples contrast from past, known ones. It was discovered four marginally unique renditions of this goal: variations location, families identification, likenesses recognition and contrasts discovery. Variations Detection. Creating variations is a standout amongst the best and least expensive techniques for an aggressor to dodge recognition systems, while reusing however much as could reasonably be expected officially accessible codes and assets. Perceiving that an example is really a variation of a known malware avoids such methodology to succeed, and makes ready to see how malware advance after some time through the improvement of new variations. Additionally this goal has been profoundly contemplated in writing, and a few evaluated papers focus on the recognition of variations. Given a noxious example m, variations location comprises in choosing from the accessible information base the examples that are variations of. Considering the colossal number of malevolent examples got day by day from significant security firms, perceiving variations of definitely known malware is pivotal to diminish the outstanding burden for human examiners.

Machine learning for malware analysis, again supplementing their commitments. quickly study writing on malware discovery and malware avoidance systems, to talk about how machine learning can be utilized by malware to sidestep current location instruments. Main review centers rather around how machine learning can bolster malware analysis, notwithstanding when avoidance strategies are utilized. Many researchers focus their overview on the location of order and control focuses through machine learning. Scientific classification of Machine Learning Techniques for Malware Analysis This area presents the scientific classification on how machine learning is utilized for malware analysis in the assessed papers. We recognize three noteworthy measurements along which studied works can be helpfully sorted out. The first describes the last target of the analysis, e.g. malware recognition. The second measurement portrays the highlights that the analysis depends on as far as how they are separated, e.g. through dynamic analysis, and what highlights are considered, e.g. CPU registers (Rath et.al, 2019). At long last, the third measurement characterizes what kind of machine learning calculation is utilized for the analysis, e.g. regulated learning. Malware Analysis Objectives Malware analysis, by and large, requests for solid recognition capacities to discover matches with the learning created by exploring past examples. Anyway, the last objective of looking for those matches contrasts. For instance, a malware expert might be explicitly keen on deciding if new suspicious examples are malevolent or not, while another might be somewhat assessing new malware searching for what family they likely have a place with. This subsection points of interest the analysis objectives of the studied papers, sorted out in three fundamental targets.

Given a deleterious example, families location comprises in choosing from the accessible information base the families that m likely has a place with. Along these lines, it is conceivable to relate obscure examples to definitely known families and, by result, give an additional esteem data to additionally examinations. Likenesses Detection. Examiners can be keen on recognizing the explicit similitudes and contrasts of the doubles to dissect as for those officially broke down. Likenesses location comprises in finding what parts and parts of an example are like something that has been as of now analyzed before. It empowers to concentrate on what is extremely new, and consequently to dispose of the rest as it doesn't merit further examination. Contrasts Detection. As a supplement, additionally distinguishing what is not the same as everything else effectively saw in the past outcomes advantageous. Actually, contrasts can direct towards finding novel viewpoints that ought to be broke down additional inside and out. Malware can be arranged by their conspicuous practices and goals. They can be keen on keeping an eye on clients' exercises and taking their touchy data (i.e., spyware), scrambling archives and requesting a payment (i.e., ransomware), or gaining remote control of a tainted machine (i.e., remote access toolboxs). Utilizing these classifications is a coarse-grained yet huge method for portraying noxious examples Although digital security firms have not as yet settled upon an institutionalized scientific categorization of malware classifications, adequately perceiving the classifications of an example can include profitable data for the analysis. The data extraction process is performed through either static or dynamic analysis, or a mix of both, while examination and relationship are completed by utilizing machine learning procedures. Methodologies dependent on static analysis take a gander at the substance of tests without requiring their execution, while dynamic analysis works by running examples to look at their conduct. A few procedures can be utilized for dynamic malware analysis. Debuggers are utilized for guidance level analysis. Test systems model and demonstrate a conduct like the earth expected by the malware, while emulators reproduce the conduct of a framework with higher precision however require more assets. Sandboxes are virtualised working frameworks giving a disconnected and solid condition where to explode malware. More nitty gritty depiction of these system are usually used to extricate highlights when dynamic analysis is utilized.

RISK ASSESSMENT IN IOT NETWORK

In light of digital ruptures, framework, vulnerabilities, assault recurrence and aggressors profile. Security risk is broke down dependent on gadget classifications and zones. Risk Mitigation is Game hypothesis based procedures that are utilized to demonstrate the risk structure. Relevant data, for example, Assessing current security levels.

There are various research oriented domains of network security in context of IoT and Machine Learning.Security Objectives/Requirements in IoT are as follows.

- System Modeling
- Identify Threats (operators and conceivable assaults)

- Identify Vulnerabilities (exploitable)
- Examining the Threat History (Likelihood)
- Counter Measures
- Risk Estimation

There are numerous functional regions of Internet of Things (IoTs) in our everyday life in which there is a high need of security and protection measures of those applications. In those applications, an assortment of IoT gadgets are utilized, for example, IoT gadgets for home and machines, lighting and warming, wellbeing checking gadgets, for example, camcorders and sensors and so forth. In wellbeing observing frameworks, gadgets for wellness, for example, wearables like FitBit Pulse, circulatory strain and glucose checking hardware and so forth. In transportation, keen answers for better transportation utilizing IoT gadgets have been created with utilization of activity flags and shrewd stopping office. In Industrial area, diverse exercises are checked utilizing IoT gadgets, for example, controlling the stream of materials, checking of oil and gas stream interferences and power use control by observing gadgets. Be that as it may, in every one of these applications security and protection have measure up to significance and are the testing issues for IoT frameworks.

As the sensor nodes in remote systems deliver high volume of information, in this way, stockpiling and their security additionally plays a major testing undertaking with Big Data related issues with in such IoT based gadgets. According to ebb and flow inquire about on Application Programming Interface, around 200 exabytes in 2014 and an estimation of 1.6 zettabytes in 2020 should be handled, 90% of these information are as of now prepared locally and the handling rate builds step by step. In a similar time the danger of basic information burglary, information and gadget control, adulteration of delicate information and also IP robbery, control and glitch of server and systems likewise can not be stayed away from. There is an extraordinary effect of information solidification and information investigation in organize setup i.e. CISCO, HPE and others. Next, in application stage regions in light of mists and firewalls at the system limits are more inclined from outer assaults.

Proactively reacting to the changing parameters of system Artificial Immune System for anchoring data frameworks dependent on human resistant framework. Fig.1. shows IoT security from research perspective. IoT needs extraordinary danger models. There is a need of conventional risk assessment structure, which can suit different danger models on equivalent terms. Most of the risk assessment philosophies are for universally useful programming frameworks and subsequently they need all encompassing methodology for evaluating risks in IoT framework because of its diversity. Also none of these location carries out risk proliferation deliberately.

CONCLUSION

Network security sphere is one of the most significant research area worked on. The Centre for Strategic and International Studies in 2014 estimated annual costs to the global economy caused by cybercrimes was between $375 billion and $575 billion. Researchers have developed some intelligent systems for network security domain with the purpose of reducing the development cost as well as to make the business network more and more secured. In this chapter newer strategies of machine learning approaches have been discussed specially ML applications of those types which can not be detected as computer programs by malware softwares/ users Artificial Intelligence based applications in view of researchers is not so easy to create an AI framework which works similar to human brain completely. Because of this, AI was started to use more specific application domain such as face recognition, object recognition etc.There is no directly contribution from human in machine learning approach . These sources of info are processing by machine learning techniques. Google isn't just self-sufficient car producer in sector. Huge numbers of the enormous companies in the vehicle business are doing research on driverless cars. For illustrative purposes, these issues were focussed that improves the network security and vulnerability.

REFERENCES

Ahmad, I., Basheri, M., Iqbal, M. J., & Rahim, A. (2018). Performance Comparison of Support Vector Machine, Random Forest, and Extreme Learning Machine for Intrusion Detection. *IEEE Access: Practical Innovations, Open Solutions*, 6, 33789–33795. doi:10.1109/ACCESS.2018.2841987

Buczak, A. L., & Guven, E. (2016). A Survey of Data Mining and Machine Learning Methods for Cyber Security Intrusion Detection. *IEEE Communications Surveys and Tutorials*, 18(2), 1153–1176. doi:10.1109/COMST.2015.2494502

Burmester, M., & de Medeiros, B. (2008). On the Security of Route Discovery in MANETs. *IEEE Transactions on Mobile Computing*, 8(9), 1180–1188.

Carvalho. (2009). Security in Mobile Ad Hoc Networks. IEEE Security and Privacy, 6(2), 72–75.

Chang, J., Tsou, P., Woungang, I., Chao, H., & Lai, C. (2015). Defending Against Collaborative Attacks by Malicious Nodes in MANETs: A Cooperative Bait Detection Approach. *IEEE Systems Journal*, 9(1), 65–75. doi:10.1109/JSYST.2013.2296197

Chaturvedi, S., Mishra, V., & Mishra, N. (2017). Sentiment analysis using machine learning for business intelligence. *IEEE International Conference on Power, Control, Signals and Instrumentation Engineering (ICPCSI)*, 2162-2166. 10.1109/ICPCSI.2017.8392100

Chen, X., Weng, J., Lu, W., Xu, J., & Weng, J. (2018). Deep Manifold Learning Combined With Convolutional Neural Networks for Action Recognition. *IEEE Transactions on Neural Networks and Learning Systems*, *29*(9), 3938–3952. doi:10.1109/TNNLS.2017.2740318 PMID:28922128

Dhurandher, Obaidat, & Verma, Gupta, & Dhurandher. (2016). FACES: Friend-Based Ad Hoc Routing Using Challenges to Establish Security in MANETs Systems. *IEEE Systems Journal*, *5*(2), 176–188.

Diro, A., & Chilamkurti, N. (2018). Leveraging LSTM Networks for Attack Detection in Fog-to-Things Communications. IEEE Communications Magazine, 56(9), 124-130. doi:10.1109/MCOM.2018.1701270

Feng, C., Wu, S., & Liu, N. (2017). A user-centric machine learning framework for cyber security operations center. *IEEE International Conference on Intelligence and Security Informatics (ISI)*, 173-175. 10.1109/ISI.2017.8004902

Ghosh & Datta. (2014). A Secure Addressing Scheme for Large-Scale Managed MANETs. *IEEE eTransactions on Network and Service Management*, *12*(3), 483–495.

He, D., Liu, C., Quek, T. Q. S., & Wang, H. (2018). Transmit Antenna Selection in MIMO Wiretap Channels: A Machine Learning Approach. *IEEE Wireless Communications Letters*, *7*(4), 634–637. doi:10.1109/LWC.2018.2805902

Islam, N., Das, S., & Chen, Y. (2017). On-Device Mobile Phone Security Exploits Machine Learning. *IEEE Pervasive Computing*, *16*(2), 92–96. doi:10.1109/MPRV.2017.26

Liu, Q., Li, P., Zhao, W., Cai, W., Yu, S., & Leung, V. C. M. (2018). A Survey on Security Threats and Defensive Techniques of Machine Learning: A Data Driven View. *IEEE Access: Practical Innovations, Open Solutions*, *6*, 12103–12117. doi:10.1109/ACCESS.2018.2805680

Mangialardo & Duarte. (2015). Integrating Static and Dynamic Malware Analysis Using Machine Learning. *IEEE Latin America Transactions, 13*(9), 3080-3087.

McDaniel, P., Papernot, N., & Celik, Z. B. (2016). Machine Learning in Adversarial Settings. *IEEE Security and Privacy*, *14*(3), 68–72. doi:10.1109/MSP.2016.51

Mozaffari-Kermani, M., Sur-Kolay, S., Raghunathan, A., & Jha, N. K. (2015). Systematic Poisoning Attacks on and Defenses for Machine Learning in Healthcare. *IEEE Journal of Biomedical and Health Informatics*, *19*(6), 1893–1905. doi:10.1109/ JBHI.2014.2344095 PMID:25095272

Nguyen, D. Q., Toulgoat, M., & Lamont, L. (2011). Impact of trust-based security association and mobility on the delay metric in MANET. *Journal of Communications and Networks (Seoul)*, *18*(1), 105–111.

Ozay, M., Esnaola, I., Yarman Vural, F. T., Kulkarni, S. R., & Poor, H. V. (2016). Machine Learning Methods for Attack Detection in the Smart Grid. *IEEE Transactions on Neural Networks and Learning Systems*, *27*(8), 1773–1786. doi:10.1109/ TNNLS.2015.2404803 PMID:25807571

Rath & Oreku. (2018). Security Issues in Mobile Devices and Mobile Adhoc Networks. In Mobile Technologies and Socio-Economic Development in Emerging Nations. IGI Global. doi:10.4018/978-1-5225-4029-8.ch009

Rath & Swain. (2018). IoT Security: A Challenge in Wireless Technology. *International Journal of Emerging Technology and Advanced Engineering, 8*(4), 43-46.

Rath & Pattanayak. (2019). Security Protocol with IDS Framework Using Mobile Agent in Robotic MANET. *International Journal of Information Security and Privacy, 13*(1), 46-58. Doi:10.4018/IJISP.2019010104

Rath, M. (2017). Resource provision and QoS support with added security for client side applications in cloud computing. *International Journal of Information Technology*, *9*(3), 1–8.

Rath, M. (2018). An Analytical Study of Security and Challenging Issues in Social Networking as an Emerging Connected Technology. *Proceedings of 3rd International Conference on Internet of Things and Connected Technologies (ICIoTCT)*.

Rath, M., & Panda, M. R. (2017). MAQ system development in mobile ad-hoc networks using mobile agents. *IEEE 2nd International Conference on Contemporary Computing and Informatics (IC3I)*, 794-798.

Rath, M., & Pati, B. (2017). *Load balanced routing scheme for MANETs with power and delay optimisation. International Journal of Communication Network and Distributed Systems* , 19.

Rath, M., & Pati, B. (2018). Security Assertion of IoT Devices Using Cloud of Things Perception. International Journal of Interdisciplinary Telecommunications and Networking, 11(2).

Rath, M., Pati, B., Panigrahi, C. R., & Sarkar, J. L. (2019). QTM: A QoS Task Monitoring System for Mobile Ad hoc Networks. In P. Sa, S. Bakshi, I. Hatzilygeroudis, & M. Sahoo (Eds.), *Recent Findings in Intelligent Computing Techniques. Advances in Intelligent Systems and Computing* (Vol. 707). Singapore: Springer. doi:10.1007/978-981-10-8639-7_57

Rath, M., Pati, B., & Pattanayak, B. (2019). Manifold Surveillance Issues in Wireless Network and the Secured Protocol. *International Journal of Information Security and Privacy, 13*(3).

Rath, M., Pati, B., & Pattanayak, B. K. (2017). Cross layer based QoS platform for multimedia transmission in MANET. *11th International Conference on Intelligent Systems and Control (ISCO)*, 402-407. 10.1109/ISCO.2017.7856026

Rath, M., & Pattanayak, B. (2017). MAQ:A Mobile Agent Based QoS Platform for MANETs. *International Journal of Business Data Communications and Networking, IGI Global, 13*(1), 1–8. doi:10.4018/IJBDCN.2017010101

Rath, M., & Pattanayak, B. (2018). Technological improvement in modern health care applications using Internet of Things (IoT) and proposal of novel health care approach. *International Journal of Human Rights in Healthcare*. doi:10.1108/IJHRH-01-2018-0007

Rath, M., & Pattanayak, B. K. (2018). Monitoring of QoS in MANET Based Real Time Applications. In Information and Communication Technology for Intelligent Systems Volume 2. ICTIS. Smart Innovation, Systems and Technologies (vol. 84, pp. 579-586). Springer. doi:10.1007/978-3-319-63645-0_64

Rath, M., & Pattanayak, B. K. (2018). SCICS: A Soft Computing Based Intelligent Communication System in VANET. Smart Secure Systems – IoT and Analytics Perspective. *Communications in Computer and Information Science, 808*, 255–261. doi:10.1007/978-981-10-7635-0_19

Rath, M., Pattanayak, B. K., & Pati, B. (2017). *Energetic Routing Protocol Design for Real-time Transmission in Mobile Ad hoc Network. In Computing and Network Sustainability, Lecture Notes in Networks and Systems* (Vol. 12). Singapore: Springer.

Rath, M., Swain, J., Pati, B., & Pattanayak, B. K. (2018). *Attacks and Control in MANET. In Handbook of Research on Network Forensics and Analysis Techniques* (pp. 19–37). IGI Global.

Rong, B., Chen, H., Qian, Y., Lu, K., Hu, R. Q., & Guizani, S. (2009). A Pyramidal Security Model for Large-Scale Group-Oriented Computing in Mobile Ad Hoc Networks: The Key Management Study. *IEEE Transactions on Vehicular Technology, 58*(1), 398–408. doi:10.1109/TVT.2008.923666

Rtah, M. (2018). Big Data and IoT-Allied Challenges Associated With Healthcare Applications in Smart and Automated Systems. *International Journal of Strategic Information Technology and Applications, 9*(2). doi:10.4018/IJSITA.201804010

Saxena, N., Tsudik, G., & Yi, J. H. (2015). Efficient Node Admission and Certificateless Secure Communication in Short-Lived MANETs. *IEEE Transactions on Parallel and Distributed Systems, 20*(2), 158–170.

Surendran & Prakash. (2014). An ACO look-ahead approach to QOS enabled fault-tolerant routing in MANETs. *China Communications, 12*(8), 93–110.

Wang, J., & Tao, Q. (2008). Machine Learning: The State of the Art. *IEEE Intelligent Systems, 23*(6), 49–55. doi:10.1109/MIS.2008.107

Wang, Yu, Tang, & Huang. (2009). A Mean Field Game Theoretic Approach for Security Enhancements in Mobile Ad hoc Networks. *IEEE Transactions on Wireless Communications, 13*(3), 1616–1627.

Wei, Z., Tang, H., Yu, F. R., Wang, M., & Mason, P. (2015). Security Enhancements for Mobile Ad Hoc Networks With Trust Management Using Uncertain Reasoning. *IEEE Transactions on Vehicular Technology, 63*(9), 4647–4658.

Xin, Y., Kong, L., Liu, Z., Chen, Y., Li, Y., Zhu, H., ... Wang, C. (2018). Machine Learning and Deep Learning Methods for Cybersecurity. *IEEE Access: Practical Innovations, Open Solutions, 6*, 35365–35381. doi:10.1109/ACCESS.2018.2836950

Chapter 4

New Tools for Cyber Security Using Blockchain Technology and Avatar–Based Management Technique

Vardan Mkrttchian
https://orcid.org/0000-0003-4871-5956
HHH University, Australia

Leyla Gamidullaeva
https://orcid.org/0000-0003-3042-7550
Penza State University, Russia

Yulia Vertakova
Southwest State University, Russia

Svetlana Panasenko
Plekhanov Russian University of Economics, Russia

ABSTRACT

The chapter introduces the perspectives on the use of avatar-based management techniques for designing new tools to improve blockchain as technology for cyber security issues. The purpose of this chapter was to develop an avatar-based closed model with strong empirical grounding that provides a uniform platform to address issues in different areas of digital economy and creating new tools to improve blockchain technology using the intelligent visualization techniques. The authors show the essence, dignity, current state, and development prospects of avatar-based management using blockchain technology for improving implementation of economic solutions in the digital economy of Russia.

DOI: 10.4018/978-1-5225-8100-0.ch004

INTRODUCTION

Management in Digital Economy is concerned with the design, execution, monitoring, and improvement of business processes. Systems that support the enactment and execution of processes have extensively been used by companies to streamline and automate intra-organizational processes. Yet, for inter-organizational processes, challenges of joint design and a lack of mutual trust have hampered a broader uptake. Emerging blockchain technology has the potential to drastically change the environment in which inter-organizational processes are able to operate. Blockchain offer a way to execute processes in a trustworthy manner even in a network without any mutual trust between nodes. Key aspects are specific algorithms that lead to consensus among the nodes and market mechanisms that motivate the nodes to progress the network. Through these capabilities, this technology has the potential to shift the discourse in management research about how systems might enable the enactment, execution, monitoring or improvement of business process within or across business networks. By using blockchain technology, untrusted parties can establish trust in the truthful execution of the code. Smart contracts can be used to implement business collaborations in general and inter-organizational business processes in particular. The potential of blockchain-based distributed ledgers to enable collaboration in open environments has been successfully tested in diverse fields ranging from diamonds trading to securities settlement *(Mendling, 2018)*.

But at this stage, it has to be noted that blockchain technology still faces numerous general technological challenges. In this chapter, we describe what we believe are the main new challenges and opportunities of blockchain technology for Digital Economy in Russia. Our study in Russia showed that the Russian research community has not addressed a majority of these challenges, albeit we note that blockchain developer communities actively discuss some of these challenges and suggest a myriad of potential solutions. Some of them can be addressed by using private or consortium blockchain instead of a fully open network. In general, the technological challenges are limited at this point, in terms of both developer support (lack of adequate tooling) and end-user support (hard to use and understand). Our recent advances on developer support include efforts by of the towards model-driven development of blockchain applications sliding mode in intellectual control and communication, help the technological challenges and created tools *(Mkrttchian and Aleshina, 2017)*.

Avatars today may communicate with each other by utilizing a variety of communications methods. This, however, has not always been the case *(Mkrttchian, 2012)*. For example, early generations of virtual worlds traditionally only supported text-based chatting features. What you "heard" another avatar "say" was really just text in a chat box within a 3-D user interface. While text- based chat is not new per se, current virtual world text-based chat is fairly sophisticated in implementation, allowing abilities such as enabling communication sender(s) to vary and tailor messages for recipient(s). For example, avatars may communicate one-on-one or one-to-many with other avatars depending on the rules of the virtual world (i.e., personal messages between avatars, zone-wide shouting, and cross-world messaging). In addition, multiple, simultaneous personalized chat channels may be utilized that can be decomposed by group membership (e.g., ad-hoc groups, teams, raids, or guilds) and governed (e.g., password protected or user-moderated chat channels). Text- based chatting in virtual worlds, like other text-based communications media, also allow for senders to edit their message prior to sending it (e.g., chat buffers) and keep extensive logs of their in-world experiences for later recall *(Mkrttchian, 2012)*. Finally, while individual keyboarding skills and network latency may be issues, text-based chat, in general, within virtual world environments is relatively quick *(Mkrttchian, 2012)*.

Based on this, management information is a specific kind of information, the management infrastructure is a kind of information infrastructure, and the subjects of management are a kind of subjects of information relations. Of particular interest is the comparison of such elements as the order (algorithm) of controlling and the system of regulation of information relations. On the one hand, in the context of the conducted element-wise comparison, they have a similar nature and purpose - they are the basis for the formation of the control effect. However, on the other hand, if the controlling order (algorithm) is a set of rules for processing information and generating control commands, then the system for regulating information relations includes not only such rules. It also includes controlling information (data on which controlling decisions are made), and governance actors, in whose role the public authorities and their officials authorized in the regulation of information relations act and the relevant infrastructure. Thus, the system of controlling of information relations is a concrete example of a cybernetic system, and information relations are the object of its control. In turn, the foregoing allows us to conclude that within the information sphere there is a cybernetic sphere (cyber sphere) - a controlling sphere, which includes a special kind of information objects, infrastructure and rules necessary for the implementation of the controlling process *(Mkrttchian, 2015b)*.

In addition to the inextricable link between the information sphere and the cyber sphere, the differences between their elements make it possible to differentiate existing threats against them, and, consequently, take these features into account while ensuring the safety of management – cyber security *(Mkrttchian, et al, 2016a)*.

If the provision of information security is an activity aimed at achieving a state of security of the information sphere in which the realization of known threats against it is impossible, and then ensuring cyber security is an activity aimed at achieving a state of management security in which its violation is impossible *(Mkrttchian, et al, 2016a)..*

In addition, the provision of information security is a category that is easily scalable; the features of the object or system within which it is implemented do not fundamentally affect its content. This activity includes ensuring the security of information, information infrastructure, actors and the system of regulating information relations both on a national scale and on a single organization scale.

In the case of cyber security, the situation is somewhat different. Its provision is already more dependent on the characteristics of the cybernetic system and management processes, since different requirements are imposed on different types of management, which must be observed.

So, for example, if it is necessary to ensure, among other things, public information about the activities of government bodies in relation to public administration, that is, its openness and availability of relevant information. In contrast, for example, the management of troops on the contrary, it is necessary to ensure its secrecy *(Mkrttchian, et al, 2016a; Mkrttchian, et al, 2016b).*

In addition, in connection with the widespread use of information technologies in the military sphere, along with the classical requirements for the management system (stability, continuity, efficiency, secrecy, efficiency), today also introduces fundamentally new requirements, such as:

- Adaptability to changing conditions and methods of using the Armed Forces;
- Providing a single information space on the battlefield;
- Openness in terms of building and capacity building;
- The possibility of reducing the operational and maintenance staff;
- Evolution in development;
- Technological independence.

Thus, the contents of the provision of cyber security can differ significantly depending on the requirements for the control system, its purpose, the specificity of the security object, the environmental conditions, the composition and state of the forces and controls, as well as the controlling order, and requires the creation of new tools for machine learning applications and Avatar-based management *(Mkrttchian, et al, 2016a; Mkrttchian, et al, 2016b).*

BACKGROUND

Modern economy networking now has become one of the most popular communication tools to have evolved over the past decade, making it a powerful new information sharing resource in world society *(Mkrttchian, et al, 2016b)*. It is known, social-economy networking is the creation and maintenance of personal and business relationships especially through online social networking service that focuses on facilitating the building of social networks or social relations among people.

By embracing social networking tools and creating standards, policies, procedures, and security measures, educational organizations can ensure that these tools are beneficial.

Academics and researchers have applied Web 2.0 technology as a way of sharing knowledge and collaborating with others in a distributed, global environment. The Internet can be considered as a research network, where knowledge is created by the all participants and shared *(Mkrttchian, et al, 2016b)*.

Currently, there are allocated some examples of ESN (Enterprise Social Networking) and services:

- Social Networks
- Social media resources
- Social Computing
- Social information processing methods
- Services of social networking
- Corporate social computing (ESC) *(Mkrttchian, et al., 2016b)*.

Lately, emerging trends and opportunities of using blockchain technology in economic sphere are considered in many works *(Swan, 2018; Hegadekatti, 2017; Morabito, 2017a, 2017b, 2017c; Calvão, 2018)*. The authors in this chapter show the essence, dignity, current state and development prospects of avatar-based management using blockchain technology for creation of new tools for machine learning applications. The purpose of this chapter is not to review the existing published work on avatar-based models for policy advice, but to try an assessment of the merits and problems of avatar-based models as a solid basis for cyber security policy advice that is mainly based on the work and experience within the recently finished projects Triple H Avatar an Avatar-based Software Platform for HHH University, Sydney, Australia which was carried out 2008-2018 *(Mkrttchian et al., 2011,2012,2013,2014, 2015, 2016, 2017, 2018)*. The agenda of this project was to develop an avatar-based closed model with strong empirical grounding and micro-foundations that provides a uniform platform to address issues in different areas of digital economy. Particular emphasis was put on the possibility to generate an implementation of the model that allows for scaling of simulation runs to large numbers of avatars tools and to provide graphical user interfaces that allow researchers not familiar with the technical details of the implementation to design (parts of) the model as well as engineering and economy experiments and to analyze simulation output *(Mkrttchian, 2015b)*.

Our study shows that blockchain technology and its application to digital economy in Russia are at an important crossroads: engineering realization issues blend with promising application scenarios; early implementations mix with unanticipated challenges. It is timely, therefore, to discuss in broad and encompassing ways where open questions lie that the scholarly community should be interested in addressing.

MAIN FOCUS OF THE CHAPTER

Issues, Controversies, Problems, Solutions and Recommendations

In this section, we discuss blockchain in relation to the visualization lifecycle including the following phases: identification, discovery, analysis, redesign, implementation, execution, monitoring, and adaptation. Using this lifecycle as a framework of reference allows us to discuss many incremental changes that blockchains might provide.

Process identification is concerned with the high-level description and evaluation of a company from a process-oriented perspective, thus connecting strategic alignment with process improvement. Currently, identification is mostly approached from an inward-looking perspective *(Dumas, et al., 2013)*. Blockchain technology adds another relevant perspective for evaluating high-level processes in terms of the implied strengths, weaknesses, opportunities, and threats. For example, how can a company systematically identify the most suitable processes for blockchains or the most threatened ones? Research is needed into how this perspective can be integrated into the identification phase. Because blockchains have affinity with the support of inter-organizational processes, process identification may need to encompass not only the needs of one organization, but broader known and even unknown partners.

Process discovery refers to the collection of information about the current way a process operates and its representation as an as-is process model. Currently, methods for process discovery are largely based on interviews, walkthroughs and documentation analysis, complemented with auto-mated process discovery techniques over non-encrypted event logs generated by process-aware information systems *(Aalst, Wil, 2016)*. Blockchain technology defines new challenges for process discovery techniques: the information may be fragmented and encrypted; accounts and keys can change frequently; and payload data may be stored partly on-chain and partly off-chain.

For example, how can a company discover an overall process from blockchain transactions when these might not be logically related to a process identifier? This fragmentation might require a repeated alignment of information from all relevant parties operating on the blockchain. Work on matching could represent a promising starting point to solve this problem *(Cayoglu, 2014; Euzenat and Shvaiko, 2013 and other)*. There is both the risk and opportunity of conducting process mining on blockchain data. An opportunity could involve establishing trust in how a process or a prospective business partner operates, while a risk is that other parties might be able to understand operational characteristics from blockchain transactions. There are also opportunities for reverse engineering business processes, among others, from smart contracts.

Process analysis refers to obtaining insights into issues relating to the way a business process currently operates. Currently, the analysis of processes mostly builds on data that is available inside of organizations or from perceptions shared by internal and external process stakeholders (Mendling, 2018). Records of processes executed on the blockchain yield valuable information that can help to assess the case load, durations, frequencies of paths, parties involved, and correlations between unencrypted data items. These pieces of information can be used to discover processes, detect deviations, and conduct root cause analysis *(Dumas, et al., 2013)*, ranging from small groups of companies to an entire industry at large. The question is which effort is required to bring the available blockchain transaction data into a format that permits such analysis. Process redesign deals with the systematic improvement of a process. Currently, approaches like redesign heuristics build on the assumption that there are recurring patterns of how a process can be improved *(Mendling, 2018)*. Blockchain technology offers novel ways of improving specific business processes or resolving specific problems. For instance, instead of involving a trustee to release a payment if an agreed condition is met, a buyer and a seller of a house might agree on a smart contract instead. The question is where blockchains can be applied for optimizing existing interactions and where new interaction patterns without a trusted central party can be established, potentially drawing on insights from related research on Web service interaction.

A promising direction for developing blockchain-appropriate abstractions and heuristics may come from data-aware workflows and diagrams. Both techniques combine two primary ingredients of blockchain, namely data and process, in a holistic manner that is well-suited for top-down design of cross-organizational processes. It might also be beneficial to formulate blockchain-specific redesign heuristics that could mimic how Incoterms define standardized interactions in international trade. Specific challenges for redesign include the joint engineering of blockchain processes between all parties involved, an ongoing problem for design (Mendling,

2018). Process implementation refers to the procedure of transforming a to-be model into software components executing the digital economy process. Currently, same processes are often implemented using process-aware information systems or business process management systems inside single organizations. In this context, the question is how the involved parties can make sure that the implementation that they deploy on the blockchain supports their process as desired. Some of the challenges regarding the transformation of a process model to blockchain artifacts are discussed by *(Mendling, 2018)*.

It has to be noted that choreographies have not been adopted by industry to a large extent yet. Despite this, they are especially helpful in inter-organizational settings, where it is not possible to control and monitor a complete process in a centralized fashion because of organizational borders. To verify that contracts between choreography stakeholders have been fulfilled, a trust basis, which is not under control of a particular party, needs to be established. Blockchains may serve to establish this kind of trust between stakeholders. An important engineering challenge on the implementation level is the identification and definition of abstractions for the design of blockchain-based business process execution. Libraries and operations for engines are required, accompanied by modeling primitives and language extensions of digital economy process. . Software patterns and anti-patterns will be of good help to engineers designing blockchain-based processes. There is also a need for new approaches for quality assurance, correctness, and verification, as well as for new corresponding correctness criteria. These can build on existing notions of compliance the more, dynamic partner binding and rebinding is a challenge that requires attention.

Process participants will have to find partners, either manually or automatically on dedicated marketplaces using dedicated look-up services. The property of inhabiting a certain role in a process might itself be a tradable asset. For example, a supplier might auction off the role of shipper to the highest bidder as part of the process. Finally, as more and more companies use blockchain, there will be a proliferation of smart contract templates available for use. Tools for finding templates appropriate for a given style of collaboration will be essential. All these characteristics emphasize the need for specific testing and verification approaches. Execution refers to the instantiation of individual cases and their information-technological processing. Currently, such execution is facilitated by process-aware information systems or digital economy process management systems. For the actual execution of a process deployed on a blockchain following the method of *(Mendling, 2018)*, several differences with the traditional ways exist.

During the execution of an instance, messages between participants need to be passed as blockchain transactions to the smart contract; resulting messages need to be observed from the blocks in the blockchain. Both of these can be achieved by integrating blockchain technology directly with existing enterprise systems or through the use of dedicated integration components, such as the triggers suggested by *(Mendling, 2018)*. The main challenge here involves ensuring correctness and security, especially when monetary assets are transferred using this technology. Process monitoring refers to collecting events of process executions, displaying them in an under-stand able way, and triggering alerts and escalation in cases where undesired behavior is observed.

Currently, such process execution data is recorded by systems that support process execution *(Mendling, 2018)*. *First*, we face issues in terms of data fragmentation and encryption as in the analysis phase. For example, the data on the blockchain alone will likely not be enough to monitor the process, but require integration with local off-chain data. Once such tracing in place, the global view of the process can be monitored independently by each involved party. This provides a suitable basis for continuous conformance and compliance checking and monitoring of service-level agreements. *Second*, based on monitoring data exchanged via the blockchain, it is possible to verify if a process instance meets the original process model and the contractual obligations of all involved process stakeholders. For this, blockchain technology can be exploited to store the process execution data and handoffs between process participants. Notably, this is even possible without the usage of smart contracts, i.e., in a first-generation blockchain like the one operated by digital economy.

Runtime adaptation refers to the concept of changing the process during execution. In traditional approaches, this can for instance be achieved by allowing participants in a process to change the model during its execution. Interacting partners might take a defensive stance in order to avoid certain types of adaptation. As discussed by *(Mendling, 2018)*, blockchain can be used to enforce conformance with the model, so that participants can rely on the joint model being followed. In such a setting, adaptation is by default something to be avoided: if a participant can change the model, this could be used to gain an unfair advantage over the other participants. For instance, the rules of retrieving digital process from an escrow account could be changed or the terms of payment. In this setting, process adaptation must strictly adhere to defined paths for it, e.g., any change to a deployed smart contract may require a transaction signed by all participants.

In contrast, the method proposed by Mendling et al. *(Mendling, 2018)* allows runtime adaptation, but assumes that relevant participants monitor the execution and react if a change is undesired. If smart contracts enforce the process, there are also problems arising in relation to evolution: new smart contracts need to be deployed to reflect changes to a new version of the process model. Porting running instances from an old version to a new one would require effective coordination mechanisms involving all participants.

There are also challenges and opportunities for digital economy process and blockchain technology beyond the classical lifecycle. We refer to the capability areas beyond the methodological support we reflected above, including strategy, governance, information technology, people, and culture. Strategic alignment refers to the active management of connections between organizational priori-ties and business processes, which aims at facilitating effective actions to improve business performance. Currently, various approaches to digital economy process assume that the corporate strategy is defined first and business processes are aligned with the respective strategic imperatives *(Mendling, 2018)*. Blockchain technology challenges these approaches to strategic alignment. For many companies, blockchains define a potential threat to their core digital economy processes. For instance, the banking industry could see a major disintermediation based on blockchain-based payment services. Also lock-in effects might deteriorate when, for example, the banking service is not the banking network itself anymore, but only the interface to it. These developments could lead to business processes and business models being under strong influence of technological innovations outside of companies.

The digital economy process governance refers to appropriate and transparent accountability in terms of roles, responsibilities, and decision processes for different digital economy-related programs, projects, and operations. Currently, digital economy processes as a management approach builds on the explicit definition of digital economy processes management-related roles and responsibilities with a focus on the internal operations of a company. Blockchain technology might change governance towards a more externally oriented model of self-governance based on smart contracts. Research on corporate governance investigates agency problems and mechanisms to provide effective incentives for intended behavior *(Mendling, 2018)*.

Smart contracts can be used to establish new governance models as exemplified by Mendling et al. (2018). It is an important question in how far this idea of Mendling can be extended towards reducing the agency problem of management discretion or eventually eliminate the need for management altogether. Furthermore, the revolutionary change suggested by Mendling, for organization shows just how disruptive this technology can be, and whether similarly radical changes could apply to digital economy processes management *(Mendling, 2018)*.

Digital economy processes management-related information technology subsumes all systems that support process execution, such as process-aware information systems and digital economy process management systems. These systems typically assume central control over the process. Blockchain technology enables novel ways of process execution, but several challenges in terms of security and privacy have to be considered. While the visibility of encrypted data on a blockchainis restricted, it is up to the participants in the process to ensure that these mechanisms are used according to their confidentiality requirements. Some of these requirements are currently being investigated in the financial industry. Further challenges can be expected with the introduction of the digital economy in Russian Federation. It is also not clear, which new attack scenarios on blockchain networks might emerge. Therefore, guidelines for using private, public, or consortium-based blockchains are required.

A person in this context refers to all individuals, possibly in different roles, who engage with digital economy processes management. Currently, these are people who work as process analyst, process manager, process owner or in other process-related roles. The roles of these individuals are shaped by skills in the area of management, business analysis and requirements engineering. In this capability area, the use of blockchain technology requires extensions of their skill sets. New required skills relate to partner and contract management, software engineering and big data analysts. Also, people have to be willing to design blockchain-based collaborations within the frame of existing regulations to enable adoption. This implies that research into blockchain-specific technology acceptance is needed, extending the established technology acceptance model *(Mendling, 2018)*.

Organizational culture is defined by the collective values of a group of people in an organization. Currently, digital economy processes management is discussed in relation to organizational culture from a perspective that emphasizes an affinity with clan and hierarchy culture. These cultural types are often found in the many companies that use digital economy processes management as an approach for documentation. Blockchains are likely to influence organizational culture towards a stronger emphasis on flexibility and an outward-looking perspective. In the competing values framework by Mendling et al. (2018), these aspects are associated with an adhocracy organizational culture. Furthermore, not only consequences of blockchain adoption have to be studied, but also antecedents'. These include organizational factors that facilitate early and successful adoption.

About Intelligent Visualization Techniques and Big Data Analytics

By itself, stored data does not generate business value, and this is true of traditional databases, data warehouses, and the new technologies for storing big data. Once the data is appropriately stored, however, it can be analyzed, which can create tremendous value. A variety of analysis technologies, approaches, and products have emerged that are especially applicable to big data, such as in-memory analytics, in-database analytics, and appliances. In our study we are using Intelligent Visualization Techniques for Big Data Analytic, or business intelligence *(Mkrttchian and Aleshina, 2017; Plotnikov, 2016; Mkrttchian, 2011; Mkrttchian, 2012 and other)*.

It is helpful to recognize that the term analytics is not used consistently; it is used in at least three different yet related ways. A starting point for understanding analytics is to explore its roots. Decision support systems (DSS) in the 1970s were the first systems to support decision making. DSS came to be used as a description for an application and an academic discipline. Over time, additional decision support applications such as executive information systems, online analytical processing (OLAP), and dashboards/scorecards became popular *(Watson and Hugh, 2014)*.

Then in the 1990s, Howard Dresner, an analyst at Gartner *(Watson and Hugh, 2014)*, popularized the term business intelligence (BI). A typical definition is that "BI is a broad category of applications, technologies, and processes for gathering, storing, accessing, and analyzing data to help business users make better decisions" (Watson and Hugh, 2014). With this definition, BI can be viewed as an umbrella term for all applications that support decision making, and this is how it is interpreted in industry and, increasingly, in academia. BI evolved from DSS, and one could argue that analytics evolved from BI (at least in terms of terminology). Thus, analytics is an umbrella term for data analysis applications. BI can also be viewed as "getting data in" (to a data mart or warehouse) and "getting data out" (analyzing the data that is stored).

A second interpretation of analytics is that it is the "getting data out" part of BI. The third interpretation is that analytics is the use of "rocket science" algorithms (e.g., machine learning, neural networks) to analyze data. These different takes on analytics do not normally cause much confusion, because the context usually makes the meaning clear *(Mkrttchian and Aleshina, 2017; Mkrttchian, 2011; Mkrttchian, 2012)*. It is useful to distinguish between three kinds of analytics because the differences have implications for the technologies and architectures used for big data analytics. Some types of analytics are better performed on some platforms than on others *(Watson and Hugh, 2014)*.

Descriptive analytics, such as reporting/OLAP, dashboards/scorecards, and data visualization, have been widely used for some time, and are the core applications of traditional BI. Descriptive analytics are backward looking (like a car's rear view mirror) and reveal what has occurred. One trend, however, is to include the findings from predictive analytics, such as forecasts of future sales, on dashboards/scorecards *(Watson and Hugh, 2014)*.

Predictive analytics suggest what will occur in the future (like looking through a car's windshield). The methods and algorithms for predictive analytics such as regression analysis, machine learning, and neural networks have existed for some time. Recently, however, software products such as SAS Enterprise Miner have made them much easier to understand and use. They have also been integrated into specific applications, such as for campaign management. Marketing is the target for many predictive analytics applications; here the goal is to better understand customers and their needs and preferences.

Some people also refer to exploratory or discovery analytics, although these are just other names for predictive analytics. When these terms are used, they normally refer to finding relationships in big data that were not previously known. The ability to analyze new data sources—that is, big data—creates additional opportunities for insights and is especially important for firms with massive amounts of customer data. Golden path analysis is a new and interesting predictive or discovery analytics technique. It involves the analysis of large quantities of behavioral data (i.e., data associated with the activities or actions of people) to identify patterns of events or activities that foretell customer actions such as not renewing a cell phone contract, closing a checking account, or abandoning an electronic shopping cart. When a company can predict a behavior, it can intercede, perhaps with an offer, and possibly change the anticipated behavior *(Watson and Hugh, 2014)*.

Whereas predictive analytics tells you what will happen, prescriptive analytics suggests what to do (like a car's GPS instructions). Prescriptive analytics can identify optimal solutions, often for the allocation of scarce resources. It, too, has been researched in *academia* for a long time but is now finding wider use in practice. For example, the use of mathematical programming for revenue management is increasingly common for organizations that have "perishable" goods such as rental cars, hotel rooms, and airline seats. For example, Harrah's Entertainment, a leader in the use of analytics, has been using revenue management for hotel room pricing for many years *(Watson and Hugh, 2014)*.

Organizations typically move from descriptive to predictive to prescriptive analytics. Another way of describing this progression is: What happened? Why did it happen? What will happen? How can we make it happen? This progression is normally seen in various BI and analytics maturity models *(Watson and Hugh, 2014)*.

There is no formula for choosing the right platforms; however, the most important considerations include the volume, velocity, and variety of data; the applications that will use the platform; that the users are; and whether the required processing is batch or real time. Some work may require the integrated use of multiple platforms. The final choices ultimately come down to where the required work can be done at the lowest cost. For our goal is good Triple H Avatar an Avatar-based Software Platform for HHH University, Sydney, Australia.

CONCLUSION

Blockchain will fundamentally shift how we deal with transactions in general, and therefore how organizations manage their business processes within their network. Discussion of challenges in relation to the digital economy processes management lifecycle and beyond points to seven major future research directions. For some of them we expect viable insights to emerge sooner, for others later. The order loosely reflects how soon such insights might appear. The digital economy processes management and the Information Systems community have a unique opportunity to help shape this fundamental shift towards a distributed, trustworthy infrastructure to promote inter-organizational processes.

ACKNOWLEDGMENT

The reported study was funded by RFBR according to the research project No. 18-010-00204_a.

REFERENCES

Belyanina, L. A. (2018). Formation of an Effective Multi-Functional Model of the Research Competence of Students. In V. Mkrttchian & L. Belyanina (Eds.), *Handbook of Research on Students' Research Competence in Modern Educational Contexts* (pp. 17–39). Hershey, PA: IGI Global. doi:10.4018/978-1-5225-3485-3.ch002

Calvão, F. (2018). Crypto-miners: Digital labor and the power of blockchain technology. *Economic Anthropology*, 6(1), 123–134. doi:10.1002ea2.12136

Cayoglu, U. (2014). *Report: The Process Model Matching Contest 2013. In BPM 2013: Business Process Management Workshops* (pp. 442–463). Springer.

Dumas, M., La Rosa, M., Mendling, J., & Reijers, H. (2013). *Fundamentals of Business Process Management*. Springer. doi:10.1007/978-3-642-33143-5

Epler, P. (2013). Using the Response to Intervention (RtI) Service Delivery Model in Middle and High Schools. *International Journal for Cross-Disciplinary Subjects in Education*, 4(1), 1089–1098. doi:10.20533/ijcdse.2042.6364.2013.0154

Euzenat, J., & Shvaiko, P. (2013). Ontology Matching: State of the Art and Future Challenges. *IEEE Transactions on Knowledge and Data Engineering*, 25(1), 158–176. doi:10.1109/TKDE.2011.253

Hegadekatti, K. (2017). *Blockchain Technology - An Instrument of Economic Evolution?* SSRN Electronic Journal; doi:10.2139srn.2943960

Khan, M. S., Ferens, K., & Kinsner, W. (2014). A Chaotic Complexity Measure for Cognitive Machine Classification of Cyber-Attacks on Computer Networks. *International Journal of Cognitive Informatics and Natural Intelligence*, 8(3), 45–69. doi:10.4018/IJCINI.2014070104

Kravets, A., Shcherbakov, M., Kultsova, M., & Shabalina, O. (2015). *Creativity in Intelligent, Technologies and Data Science – 2015*. First Conference, CIT&DS 2015, Volgograd, Russia. doi: 10.1007/978-3-319-23766-4

Mendling, J. (2018). Blockchains for Business Process Management – Challenges and Opportunities. ACM Trans. Manag. Inform. Syst., 9.

Mkrttchian, V. (2011). Use 'hhh" technology in transformative models of online education. In G. Kurubacak & T. Vokan Yuzer (Eds.), *Handbook of research on transformative online education and liberation: Models for social equality* (pp. 340–351). Hershey, PA: IGI Global. doi:10.4018/978-1-60960-046-4.ch018

Mkrttchian, V. (2012). Avatar manager and student reflective conversations as the base for describing meta-communication model. In G. Kurubacak, T. Vokan Yuzer, & U. Demiray (Eds.), *Meta-communication for reflective online conversations: Models for distance education* (pp. 340–351). Hershey, PA: IGI Global. doi:10.4018/978-1-61350-071-2.ch005

Mkrttchian, V. (2015a), Use Online Multi-Cloud Platform Lab with Intellectual Agents: Avatars for Study of Knowledge Visualization & Probability Theory in Bioinformatics. International Journal of Knowledge Discovery in Bioinformatics, 5(1), 11-23. Doi:10.4018/IJKDB.2015010102

Mkrttchian, V. (2015b). Modeling using of Triple H-Avatar Technology in online Multi-Cloud Platform Lab. In M. Khosrow-Pour (Ed.), Encyclopedia of Information Science and Technology (3rd Ed.). (pp. 4162-4170). Hershey, PA: IGI Global. Doi:10.4018/978-1-4666-5888-2.ch409

Mkrttchian, V., & Aleshina, E. (2017). *Sliding Mode in Intellectual Control and Communication: Emerging Research and Opportunities*. Hershey, PA: IGI Global; doi:10.4018/978-1-5225-2292-8

Mkrttchian, V., & Aleshina, E. (2017). The Sliding Mode Technique and Technology (SM T&T) According to VardanMkrttchian in Intellectual Control(IC). In *Sliding Mode in Intellectual Control and Communication: Emerging Research and Opportunities* (pp. 1–9). Hershey, PA: IGI Global. doi:10.4018/978-1-5225-2292-8.ch001

Mkrttchian, V., Bershadsky, A., Bozhday, A., & Fionova, L. (2015). Model in SM of DEE Based on Service Oriented Interactions at Dynamic Software Product Lines. In G. Eby & T. Vokan Yuzer (Eds.), *Identification, Evaluation, and Perceptions of Distance Education Experts* (pp. 230–247). Hershey, PA: IGI Global. doi:10.4018/978-1-4666-8119-4.ch014

Mkrttchian, V., Bershadsky, A., Bozhday, A., Kataev, M., & Kataev, S. (Eds.). (2016b). *Handbook of Research on Estimation and Control Techniques in E-Learning systems*. Hershey, PA: IGI Global. doi:10.4018/978-1-4666-9489-7

Mkrttchian, V., Bershadsky, A., Bozhday, A., Noskova, T., & Miminova, S. (2016a). Development of a Global Policy of All-Pervading E-Learning, Based on Transparency, Strategy, and Model of Cyber Triple H-Avatar. In G. Eby, T. V. Yuser, & S. Atay (Eds.), *Developing Successful Strategies for Global Policies and Cyber Transparency in E-Learning* (pp. 207–221). Hershey, PA: IGI Global. doi:10.4018/978-1-4666-8844-5.ch013

Mkrttchian, V., Kataev, M., Hwang, W., Bedi, S., & Fedotova, A. (2014). Using Plug-Avatars "hhh" Technology Education as Service-Oriented Virtual Learning Environment in Sliding Mode. In G. Eby & T. Vokan Yuzer (Eds.), *Emerging Priorities and Trends in Distance Education: Communication, Pedagogy, and Technology*. Hershey, PA: IGI Global. doi:10.4018/978-1-4666-5162-3.ch004

Mkrttchian, V., Kataev, M., Hwang, W., Bedi, S., & Fedotova, A. (2016), Using Plug-Avatars "hhh" Technology Education as Service-Oriented Virtual Learning Environment in Sliding Mode. In Leadership and Personnel Management: Concepts, Methodologies, Tools, and Applications (pp. 890-902). IGI Global. Doi:10.4018/978-1-4666-9624-2.ch039

Mkrttchian, V., Palatkin, I., Gamidullaeva, L. A., & Panasenko, S. (2019). About Digital Avatars for Control Systems Using Big Data and Knowledge Sharing in Virtual Industries. In A. Gyamfi & I. Williams (Eds.), *Big Data and Knowledge Sharing in Virtual Organizations* (pp. 103–116). Hershey, PA: IGI Global. doi:10.4018/978-1-5225-7519-1.ch004

Morabito, V. (2017a). Blockchain and Enterprise Systems. *Business Innovation Through Blockchain*, 125–142. doi:10.1007/978-3-319-48478-5_7

Morabito, V. (2017b). The Blockchain Paradigm Change Structure. *Business Innovation Through Blockchain,* 3–20. doi:10.1007/978-3-319-48478-5_1

Morabito, V. (2017c). Blockchain Practices. *Business Innovation through Blockchain,* 145–166. doi:10.1007/978-3-319-48478-5_8

Plotnikov, V., Vertakova, Y., & Leontyev, E. (2016). Evaluation of the effectiveness of the telecommunication company's cluster management. *Economic Computation and Economic Cybernetics Studies and Research,* 50(4), 109–118.

Swan, M. (2018). Blockchain Economic Networks: Economic Network Theory—Systemic Risk and Blockchain Technology. *Business Transformation through Blockchain,* 3–45. doi:10.1007/978-3-319-98911-2_1

Tran, T. P., Tsai, P., Jan, T., & He, X. (2012). Machine Learning Techniques for Network Intrusion Detection. In Machine Learning: Concepts, Methodologies, Tools and Applications (pp. 498-521). Hershey, PA: IGI Global. doi:10.4018/978-1-60960-818-7.ch310

Van der Aalst, W. M. P. (2016). *Process Mining: Data Science in Action.* Springer. doi:10.1007/978-3-662-49851-4

Watson, H. J. (2014). Tutorial: Big Data Analytics: Concepts, Technologies, and Applications. *Communications of the Association for Information Systems, 34,* 65. doi:10.17705/1CAIS.03465

ADDITIONAL READING

Mkrttchian, V. (2016). The Control of Didactics of Online Training of Teachers in HHH University and Cooperation with the Ministry of Diaspora of Armenia. In V. Mkrttchian, A. Bershadsky, A. Bozhday, M. Kataev, & S. Kataev (Eds.), *Handbook of Research on Estimation and Control Techniques in E-learning systems* (pp. 311–322). Hershey, PA, USA: IGI Global; doi:10.4018/978-1-4666-9489-7.ch021

Mkrttchian, V., & Belyanina, L. (2016). The Pedagogical and Engineering Features of E- and Blended Learning of Aduits Using Triple H-Avatar in Russian Federation. In V. Mkrttchian, A. Bershadsky, A. Bozhday, M. Kataev, & S. Kataev (Eds.), *Handbook of Research on Estimation and Control Techniques in E-Learning Systems* (pp. 61–77). Hershey, PA, USA: IGI Global; doi:10.4018/978-1-4666-9489-7.ch006

KEY TERMS AND DEFINITIONS

Audit and Policy Mechanisms: Is section of avatar-based management techniques.

Avatar-Based Management: Is a control method and technique introduced by V. Mkrttchian in 2018.

Blockchain: Is a growing list of records, called blocks, which are linked using cryptography. Each block contains a cryptographic hash of the previous block a timestamp, and transaction data.

Cyber Security: Is section of information security, within the framework of which the processes of formation, functioning and evolution of cyber objects are studied, to identify sources of cyber-danger formed while determining their characteristics, as well as their classification and formation of regulatory documents, implementation of security systems in future.

Machine Learning Application: Is a class of methods of artificial/natural intelligence, the characteristic feature of which is not a direct solution of the problem, but training in the process of applying solutions to a set of similar problems.

Maturity Models: Is section of avatar-based management.

Predictive Analytics: Is the use of data, statistical algorithms and machine learning techniques to identify the likelihood of future outcomes based on empirical data.

Chapter 5
Machine Learning Application With Avatar-Based Management Security to Reduce Cyber Threat

Vardan Mkrttchian
https://orcid.org/0000-0003-4871-5956
HHH University, Australia

Leyla Gamidullaeva
https://orcid.org/0000-0003-3042-7550
Penza State University, Russia & K. G. Razumovsky Moscow State University of Technologies and Management, Russia

Yulia Vertakova
Southwest State University, Russia

Svetlana Panasenko
Plekhanov Russian University of Economics, Russia

ABSTRACT

This chapter is devoted to studying the opportunities of machine learning with avatar-based management techniques aimed at optimizing threat for cyber security professionals. The authors of the chapter developed a triangular scheme of machine learning, which included at each vertex one participant: a trainee, training, and an expert. To realize the goal set by the authors, an intelligent agent is included in the triangular scheme. The authors developed the innovation tools using intelligent visualization techniques for big data analytic with avatar-based management in sliding mode introduced by V. Mkrttchian in his books and chapters published by IGI Global in 2017-18. The developed algorithm, in contrast to the well-known,

DOI: 10.4018/978-1-5225-8100-0.ch005

uses a three-loop feedback system that regulates the current state of the program depending on the user's actions, virtual state, and the status of implementation of available hardware resources. The algorithm of automatic situational selection of interactive software component configuration in virtual machine learning environment in intelligent-analytic platforms was developed.

INTRODUCTION

Existing security systems offer a reasonable level of protection; however, they cannot cope with the growing complexity of computer networks and hacking techniques. Moreover, security systems suffer from low detection rates and high false alarm rates. In order to overcome such challenging problems, there has been a great number of research conducted to apply Machine Learning (ML) algorithms (Tran, et al., 2012). Machine learning techniques have been successfully applied to several real world problems in areas as diverse as image analysis, Semantic Web, bioinformatics, text processing, natural language processing, telecommunications, finance, medical diagnosis, and so forth (Gama, and Carvalho, 2012).

Recent definition of machine learning is developed by I. Cadez, P. Smyth, H. Mannila, A. Salah, E. Alpaydin (*Cadez, et al., 2001; Salah and Alpaydin, 2004*). The issues of the use of machine learning in cyber security are disclosed in many works (*Anagnostopoulos, 2018; Edgar and Manz, 2017; Yavanoglu and Aydos, 2017; Khan, et al., 2014; Khan, 2019; Dinur, 2018*). Using data mining and machine learning methods for cyber security intrusion detection is proposed by the authors. (Kumar, et al., 2017)

Object classification literature shows that computer software and hardware algorithms are increasingly showing signs of cognition and are necessarily evolving towards cognitive computing machines to meet the challenges of engineering problems (*Khan, et al, 2014*). For instance, in response to the continual mutating nature of cyber security threats, basic algorithms for intrusion detection are being forced to evolve and develop into autonomous and adaptive agents, in a manner that is emulative of human information processing mechanisms and processes (*Khan, et al., 2014; Khan, 2019*).

In connection with the widespread use of information technologies in the military and state fields, along with the classical requirements for the controlling system (stability, continuity, efficiency, secrecy, efficiency), today also introduces fundamentally new requirements, such as:

- adaptability to changing conditions and methods of using the Armed Forces and State;
- providing a single information space on the battlefield;
- openness in terms of building and capacity building;
- possibility of reducing the operational and maintenance staff;
- evolution in development;
- technological independence.

Thus, the maintenance of cyber security can significantly differ depending on the requirements for the control system, its purpose, the specificity of the managed object, the environmental conditions, the composition and state of the forces and controls, and the management order. Why do we need to distinguish between information and cyber security? What tasks can be achieved with this distinction? This need is conditioned by the transition to a new socio-economic formation, called the information society.

If earlier the problems of ensuring cyber security were relevant mainly for the military organization, in connection with the existence and development of the forces and means of information confrontation and electronic warfare, now such problems exist for the state as a whole.

Among the reasons for this situation can be called:

- The absence of an international legal basis prohibiting the use of information weapons and conducting information operations;
- Imperfection of the regulatory legal framework establishing liability for the commission of crimes in the field of information technology;
- Development by individual states of doctrines and strategies of offensive and subversive actions in the information space;
- Intensive development of military information technologies, including means of destruction of civil and military control systems;
- Leveling the role of international organizations and their bodies in the field of ensuring international information security;
- Creation and use of Special Forces and means of negative impact on critical information infrastructure;
- The existence of special samples of malicious software affecting the automated control systems of industrial and other objects of critical infrastructure;
- The emergence of forms of civil disobedience associated with encroachments on the information infrastructure in protest against the policy of the state and the activities of government bodies;

- Penetration of information technologies in all spheres of state and public life, building on their basis systems of state and military control;
- Development of state projects and programs in the field of information (electronic document management, interagency electronic interaction, universal electronic cards, provision of public services in electronic form) aimed at the formation of an information society;

Thus, the tasks of ensuring cyber security for today exist, both for the state as a whole, and for certain critical structures, systems and objects.

Let us examine in more detail some of these reasons that necessitate the isolation of cyber security as an independent type of security.

One of the new negative phenomena that pose a threat to the information sphere is the emergence of hacker groups taking an active social position and covering their activities in social networks and the media. Unlike classic hacker groups that seek not only to hide their activities, but also the fact of their existence, so-called hack activists position themselves as fighters against injustice and arbitrariness, and follow certain independently developed rules and principles. Moreover, the analysis of the activities of such groups allows us to conclude that they pursue non-commercial goals. In their work, ideological (political), rather than material motivation, prevails, which turns them into a more serious threat to the state and commercial companies, compared to ordinary criminals. The analytical study as one of the tendencies of information security in now indicated an increase in attacks on top management. Top executives are no longer visible on the web. Firms should be allowed the possibility that hackers already have complete information about their leadership, which can be used both to discredit and damage the reputation of the company, and for targeted attacks. These facts allow us to conclude that part of the hacker community has become interested in the political and economic situation and is trying to influence it by organizing and holding mass protests and large-scale hacker attacks on the resources of state, including military and commercial structures. The existence of such groups, their ideology and political motivation, negative attitude of their members to public authorities and disregard for the established law and order, present a real danger to the state, including military administration and its infrastructure.

Many experts on information security call hack activism one of the main negative trends of the past and the coming years. Given the fact that the activities of such groups are of a Tran's boundary nature and are expressed in the conduct of unlawful actions in the information space, it is also a threat to the information spheres around the world.

BACKGROUND

Another serious threat for cyber security professionals is the emergence and development of malicious software hitting automated control systems. By embracing social networking tools and creating standards, policies, procedures, and security measures, educational organizations can ensure that these tools are beneficial. At present, many world organizations and enterprises underwent a virus attack, information about which has not been disclosed for security reasons. These incidents attracted the closest attention of security experts around the world. As a result of numerous studies, vulnerabilities have been found in almost all cyber security systems. Much vulnerability allows attackers to remotely execute code in the most important systems responsible for monitoring equipment and receiving inbound information. At the same time, the authors of the chapter believe that the creation of such complex malicious programs are highly skilled personnel and large financial resources, and further appearance of new similar samples should be expected, therefore, modern cybernetic safety must be built as a multidisciplinary science at the highest level of professionalism. The authors of this chapter, as specialists in intellectual control and communications in sliding mode, offer their own, original approach to solving the problem based on the author's method - machine learning applications with avatar-based management techniques. Thus, the described facts make one look at the security of industrial facilities, including defense enterprises, in a new perspective. Along with classical security measures, now it is necessary to pay attention to the security of control systems, to identify and eliminate possible vulnerabilities in their components and infrastructure.

Another example, which makes it necessary to separate cyber security as a separate category, is the activity of foreign countries in the creation of special units and structures. The so-called centers of cyber defense (defense) have appeared in a number of countries lately. Such centers in one form or another already exist in the United States, UK, Australia, Israel, Iran, China, and Germany.

In general, the tasks of such units in all countries are approximately the same - together with special services and law enforcement bodies they protect government bodies, as well as civilian and military critical facilities from harmful effects (hacker attacks, malicious software, etc.). However, their very name, purpose, main tasks and affiliation, in most cases, to a military organization, testify to the priority of the goals of ensuring the security of state and military management. As our research has shown for information confrontation, special information and strike groups of forces and means are created, the objectives of which are as a rule: neutralization or destruction of the information and strategic resource of another (opposing) entity and its armed forces and ensuring the protection of its information and strategic resource from the similar impact of the enemy.

MAIN FOCUS OF THE CHAPTER

Issues, Controversies, Problems, Solutions and Recommendations

The main objects of impact for these forces and assets are:

- Software and information support;
- Software and hardware, telecommunications and other means of information and management;
- Communication channels that ensure the circulation of information flows and the integration of the management system;
- Human intellect and mass consciousness.

Emphasizing the special importance of the information sphere and related processes for the defense of the state, this author identifies military information security as one of the types of military security.

Recently, the attitude towards the information confrontation has changed significantly, it is no longer perceived only as an activity accompanying military operations. More and more factors are forced to treat him as an independent and most promising type of negative impact on the enemy.

In its new quality, such an impact is not only outside the framework of classical military operations, but also gets wider content.

It becomes a powerful, invisible and undeclared by international law means, allowing in the shortest possible time to paralyze the main forces and assets of a hostile state without causing fatal damage to its industrial facilities and territory.

That is why it is necessary to further develop the corresponding developments of military science and their application not only within the framework of the military organization, but also the entire information sphere of any state. The current situation and the specifics of new threats blur the boundaries between peaceful and war time and require the constant participation of all the forces and means of the state in ensuring its information security.

To achieve these goals, complete consolidation of efforts and a clear interaction of law enforcement agencies, special services, military units, research organizations, the media, public associations and commercial companies are necessary.

The Avatar-Based Management technique developed by V. Mkrtchian for the use in modern education process was introduced in journals and books published by IGI Global in 2014-2018 *(Mkrttchian, et al, 2014; Mkrttchian, 2015; Mkrttchian, et al, 2016; Mkrttchian and Aleshina, 2017; Mkrttchian and Belyanina, 2018)*. For the purpose of breaking down language and cultural barriers in modern Russian corporations, V. Mkrttchian proposed to use the technology and methods of Avatar-Based Management for training and the improved RTI technology - Respond to Intervention, successfully used in the US in the special education system. Response to Intervention has been in existence for only a short period, yet it has had a powerful impact on the academic achievement of students across the United States. The National Center on Response to Intervention (2010) defined Response to Intervention as a delivery service model that integrates assessment and intervention within a multi-level prevention system to maximize student achievement and reduce behavior problems. With Response to Intervention, schools identify students at risk for poor learning outcomes, monitor student progress, provide evidence-based interventions and adjust the intensity and nature of those interventions depending on a student's responsiveness, and identify students with learning disabilities or other disabilities. To solve the problems with the Russian language, they propose using the previously obtained teaching technologies together with the RTI technology for a new field of application - training the Russian language for personnel of state corporations *(Vertakova and Plotnikov, 2014; Mkrttchian, 2017; Epler, 2013)*. The authors of the chapter developed the innovation tools, based on the analysis results of the deliberations of citizens in social networks on topics related to online services, using Intelligent Visualization Techniques for Big Data Analytics with Avatar-Based Management Techniques *(Mkrttchian, et al, 2019)*. Social networks and the blogosphere, which is a popular and active area of mass communication, may become the subject of study and data source for the demand for e-government services. Social networks enjoy great popularity among Russian citizens. According to com. Score, nearly every Russian Internet user (99.7% of the average domestic Internet users) has an account in social networks. Scale users of Russian social networks account for more than 52 million people. The average Russian social network user spends 12, 8 hours per month on social networks, which is the highest rate in the world (the world average is 5.9 hours per month) (source and compare). Over the past decade, the social network came on the audience of the Central Russian TV channels. These facts highlight the demand and popularity of social media in Russia, as well as the possibility of using the discussions in social networks to identify attitudes, opinions of citizens and their assessment of the activities of the state.

The problem of discrepancy between the legal basis, both international and domestic, of modern information relations deserves special attention; its inability to take into account the danger of new negative phenomena threatening the information sphere in general and its separate elements, but this problem is beyond the scope of this chapter and will be considered separately by the authors. World policy for cyber security advice in the World Economy requires a thorough understanding of the relevant economic mechanisms that are responsible for the (overall) effects of policy measures in the digital economy *(Mkrtchian, et al., 2019)*. As they do not see a model which reflects their perception of the world advice based on highly abstract vehicles of thinking is likely to be rejected. Avatar-based models may be less prone to be rejected by policymakers as they usually are characterized by a lot more economic structure. This is not to say that an avatar-based modeler would choose any other general approach of building his model than a more orthodox economist. It is rather the larger toolbox that avatar-based models offer which allows him to bring into the picture features of the system that policymakers may find more convincing. Avatars can for example be endowed with different behavioral rules which policy makers recognize from own experience. It is feasible to model an economy along its spatial dimension, and institutions can be incorporated in a much more fine-grained way as in more traditional approaches. As the policy-maker's part is usually about deciding on the institutional environment and possible changes of that, having a more accessible model in that respect may be of great value for a fruitful interaction between policy advisers and policymakers. While most of our discussion so far focused on how to write down an avatar-based model that brings into the picture a simultaneous analysis of various non-negligible institutional, spatial or economic features for a better policymaking, an underdeveloped branch of avatar-based modeling is certainly the positive analysis of economic policy making. It occurs to us that avatar-based models are far from being fully exploited as a means of positive policy analysis. More should be done to bring together a meaningful economic model with an equally meaningful political model that does justice to the intricate rules which characterize democratic societies and shape policy outcomes. Admittedly most of our selling points had the flavor of "we – the avatar-based modelers - can do more". This should not be misunderstood as an argument that in general bigger models are better. Quite on the contrary, it seems crucial to us, that, regardless whether analytical or simulation methods are employed, models are carefully built in a way that only those aspects of the economic environment which seem directly relevant for the policy question at hand are modeled in some detail. A closed macroeconomic avatar-based model has to contain all relevant market, but this does by no means imply that all these markets have to be modeled with identical granularity and institutional richness. Nevertheless, closed macroeconomic avatar-based models typically are quite large and building big models requires big computing power. But machines that potentially can do the job exist and are used by other professions like meteorologists or physicists. However, a lesson learned from our Triple H Avatar project was that running economic models on parallel machines brings up new and non-trivial problems. The reason behind these technical issues is quite intuitive: parallelization requires the slicing up of a big task into digestible smaller chunks. The question becomes how to cut through an economic system. This chapter goal is about machine learning with avatar-based management security; use to optimize threat for cyber security professionals. The chapter illustrates results developed new triangular scheme of machine learning, which included at each vertex one participant: a trainee, training and an expert. Realize the goal set by the authors of the chapter, an intelligent agent is included in the triangular scheme, developed the innovation tools, using Intelligent Visualization Techniques for Big Data Analytic with Avatar-Based Management Techniques. Developed algorithm, in contrast to the well-known, uses three-loop feedback system, that regulate the current state of the program, depending on the user's actions, virtual state and the status of implementation of available hardware resources. Algorithm of automatic situational selection of interactive software component configuration in virtual machine learning environment in intelligent-analytic platforms was developed see fig.1 and fig.2).

An obvious candidate is the spatial dimension of an economic model, i.e. to allocate the computing to be done for a particular region to a particular processor. However, as there is considerable interaction between regions as in economic models factor and product markets are typically highly interdependent across regions via the flows of worker, capital, intermediate or final goods, a lot of communication between processors has to be organized which can considerable slow down the computing. In order to be able to use avatar-based models for economic policy advice in a way we sketched it, problems of parallelizing code or in general computing issues need to be resolved. In addition and coming back to our argument of convincing policymakers of the appropriateness of the framework on which the policy advice is based, easy to

Figure 1. Triangular scheme of virtual machine learning environment

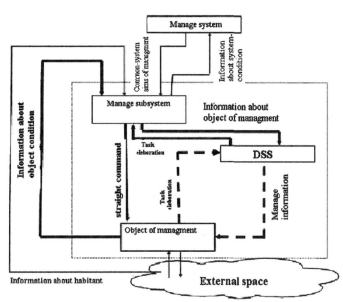

Figure 2. Algorithm of Self-organizing intelligent and analytics platforms

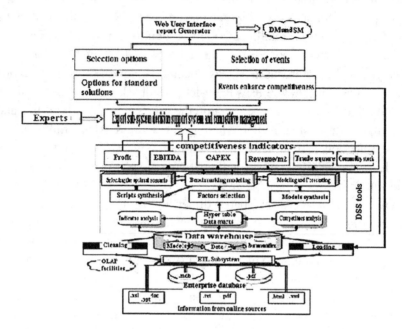

use and intuitive graphical user interfaces (GUIs) need to be developed. Ideally, at some point these GUIs would be so user-friendly that any interested person would be able to run his own simulations. Another issue down the road, which we find important as we want to precede to using avatar-based models for economic policy advice, is the behavioral foundations of the (heterogeneous) avatars that populate our models. Once we deviate or even abandon the perfectly rational avatar model of High Technology there are many degrees of freedom on what to assume for the behavior of an avatar, may it be a worker, firms or a government agency. In the Triple H Avatar project we followed the modeling philosophy to apply management rules for modeling firm behavior. For most decision problems firms face the management literature offers standard procedures (which are often heuristic methods). Examples are specifications on how firms plan their production volume or replenish their stocks. Some of these suggestions are even implemented in standard software that is purchased by firms to automatize on these operational management decisions. As we want to base the policy advice on models where firm behavior is as close as possible to the performance of real world firms it seems natural and also for outsiders convincing to rely on such standard rules where available. For the modeling of the behavior of individuals a promising approach seems to be incorporate findings from experimental studies.

CONCLUSION

In the present conditions, cybernetic security, as a scientific and legal category, exists not only within the framework of information warfare and military security, but also the information sphere as a whole, and therefore should be viewed as an independent type of security. The maintenance of cyber security can significantly differ depending on the requirements for the control system, its purpose, the specificity of the managed object, the environmental conditions, the composition and state of the forces and controls, and the management order.

In conclusion, the authors came to the following research results:

- **Cyber Security:** The state of security management, in which its violation is impossible.
- **Ensuring Cyber Security:** Activities aimed at achieving a state of security management, in which its violation is impossible.
- **Cybernetic Threats:** Phenomena, acts, conditions, factors that pose a threat to management information, management infrastructure, governance entities and governance. The danger lies in the possibility of disrupting the properties of one or more of these elements, which can lead to disruption of control.
- **Cybernetic Attack:** Actions aimed at violation of management.
- The isolation of cyber security as an independent type of security is due to the current level of development of the information sphere, it is justified and allows to fully realize the potential of legal, organizational and technical protection measures, taking into account the specifics of the corresponding systems and management processes. In turn, this implies an increase in the requirements for the protection of systems and management processes, depending on their level, composition and purpose.
- In order to increase the legal protection of cybernetic systems of critical facilities, it seems necessary to expand the list of crimes in the field of computer information and to establish a more severe criminal liability for infringement of such objects.

ACKNOWLEDGMENT

The reported study was funded by RFBR according to the research project No. 18-010-00204_a.

REFERENCES

Anagnostopoulos, C. (2018). Weakly Supervised Learning: How to Engineer Labels for Machine Learning in Cyber-Security. *Data Science for Cyber-Security*, 195–226. doi:10.1142/9781786345646_010

Belyanina, L. A. (2018). Formation of an Effective Multi-Functional Model of the Research Competence of Students. In V. Mkrttchian & L. Belyanina (Eds.), *Handbook of Research on Students' Research Competence in Modern Educational Contexts* (pp. 17–39). Hershey, PA: IGI Global. doi:10.4018/978-1-5225-3485-3.ch002

Cadez, I. V., Smyth, P., & Mannila, H. (2001). Probabilistic modeling of transaction data with applications to profiling, visualization, and prediction. *Proceedings of the Seventh ACM SIGKDD International Conference on Knowledge Discovery and Data Mining - KDD '01*. doi:10.1145/502512.502523

Dinur, I., Dolev, S., & Lodha, S. (Eds.). (2018). Lecture Notes in Computer Science Cyber Security Cryptography and Machine Learning. Springer. doi:10.1007/978-3-319-94147-9

Edgar, T. W., & Manz, D. O. (2017). Machine Learning. *Research Methods for Cyber Security*, 153–173. doi:10.1016/b978-0-12-805349-2.00006-6

Epler, P. (2013). Using the Response to Intervention (RtI) Service Delivery Model in Middle and High Schools. *International Journal for Cross-Disciplinary Subjects in Education*, 4(1), 1089–1098. doi:10.20533/ijcdse.2042.6364.2013.0154

Gama, J., & de Carvalho, A. C. (2012). Machine Learning. In I. Management Association (Ed.), Machine Learning: Concepts, Methodologies, Tools and Applications (pp. 13-22). Hershey, PA: IGI Global. doi:10.4018/978-1-60960-818-7.ch102

Khan, M. S. (Ed.). (2019). Machine Learning and Cognitive Science Applications in Cyber Security. Academic Press. doi:10.4018/978-1-5225-8100-0

Khan, M. S., Ferens, K., & Kinsner, W. (2014). A Chaotic Complexity Measure for Cognitive Machine Classification of Cyber-Attacks on Computer Networks. *International Journal of Cognitive Informatics and Natural Intelligence*, 8(3), 45–69. doi:10.4018/IJCINI.2014070104

Kumar, R., & Rituraj, S. (2017). Using Data Mining and Machine Learning Methods for Cyber Security Intrusion Detection. *International Journal of Recent Trends in Engineering and Research*, 3(4), 109–111. doi:10.23883/IJRTER.2017.3117.9NWQV

Mkrttchian, V. (2015). Use Online Multi-Cloud Platform Lab with Intellectual Agents: Avatars for Study of Knowledge Visualization & Probability Theory in Bioinformatics. International Journal of Knowledge Discovery in Bioinformatics, 5(1), 11-23. Doi:10.4018/IJKDB.2015010102

Mkrttchian, V. (2017). Project-Based Learning for Students with Intellectual Disabilities. In P. L. Epler (Ed.), *Instructional Strategies in General Education and Putting the Individuals With Disabilities Act (IDEA) Into Practice* (pp. 196–221). Hershey, PA: IGI Global. doi:10.4018/978-1-5225-3111-1.ch007

Mkrttchian, V., & Aleshina, E. (2017). *Sliding Mode in Intellectual Control and Communication: Emerging Research and Opportunities*. Hershey, PA: IGI Global. doi:10.4018/978-1-5225-2292-8

Mkrttchian, V., & Belyanina, L. (Eds.). (2018). *Handbook of Research on Students' Research Competence in Modern Educational Contexts*. Hershey, PA: IGI Global. doi:10.4018/978-1-5225-3485-3

Mkrttchian, V., Bershadsky, A., Bozhday, A., Kataev, M., & Kataev, S. (Eds.). (2016). *Handbook of Research on Estimation and Control Techniques in E-Learning systems*. Hershey, PA: IGI Global. doi:10.4018/978-1-4666-9489-7

Mkrttchian, V., Kataev, M., Shih, T., Kumar, M., & Fedotova, A. (2014). Avatars "HHH" Technology Education Cloud Platform on Sliding Mode Based Plug- Ontology as a Gateway to Improvement of Feedback Control Online Society. International Journal of Information Communication Technologies and Human Development, 6(3), 13-31. Doi:10.4018/ijicthd.2014070102

Mkrttchian, V., Palatkin, I., Gamidullaeva, L. A., & Panasenko, S. (2019). About Digital Avatars for Control Systems Using Big Data and Knowledge Sharing in Virtual Industries. In A. Gyamfi & I. Williams (Eds.), *Big Data and Knowledge Sharing in Virtual Organizations* (pp. 103–116). Hershey, PA: IGI Global. doi:10.4018/978-1-5225-7519-1.ch004

Salah, A. A., & Alpaydin, E. (2004). Incremental mixtures of factor analysers. *Proceedings of the 17th International Conference on Pattern Recognition*. 10.1109/ICPR.2004.1334106

Tran, T. P., Tsai, P., Jan, T., & He, X. (2012). Machine Learning Techniques for Network Intrusion Detection. In I. Management Association (Ed.), Machine Learning: Concepts, Methodologies, Tools and Applications (pp. 498-521). Hershey, PA: IGI Global. doi:10.4018/978-1-60960-818-7.ch310

Vertakova, J., & Plotnikov, V. (2014). Public-private partnerships and the specifics of their implementation in vocational education. *Proceeded Economics and Finance*, *16*, 24–33. doi:10.1016/S2212-5671(14)00770-9

Yavanoglu, O., & Aydos, M. (2017). A review on cyber security datasets for machine learning algorithms. *2017 IEEE International Conference on Big Data (Big Data)*. 10.1109/BigData.2017.8258167

ADDITIONAL READING

Mkrttchian, V. (2012). Avatar manager and student reflective conversations as the base for describing meta-communication model. In G. Kurubacak, T. Vokan Yuzer, & U. Demiray (Eds.), *Meta-communication for reflective online conversations: Models for distance education* (pp. 340–351). Hershey, PA, USA: IGI Global; doi:10.4018/978-1-61350-071-2.ch005

Mkrttchian, V. (2015). Modeling using of Triple H-Avatar Technology in online Multi-Cloud Platform Lab. In M. Khosrow-Pour (Ed.), Encyclopedia of Information Science and Technology (3rd Ed.). (pp. 4162-4170). IRMA, Hershey: PA, USA: IGI Global. Doi:10.4018/978-1-4666-5888-2.ch409

Mkrttchian, V., & Aleshina, E. (2017). The Sliding Mode Technique and Technology (SM T&T) According to Vardan Mkrttchian in Intellectual Control(IC). In *Sliding Mode in Intellectual Control and Communication: Emerging Research and Opportunities* (pp. 1–9). Hershey, PA: IGI Global; doi:10.4018/978-1-5225-2292-8.ch001

Mkrttchian, V., Amirov, D., & Belyanina, L. (2017). Optimizing an Online Learning Course Using Automatic Curating in Sliding Mode. In N. Ostashewski, J. Howell, & M. Cleveland-Innes (Eds.), *Optimizing K-12 Education through Online and Blended Learning* (pp. 213–224). Hershey, PA, USA: IGI Global; doi:10.4018/978-1-5225-0507-5.ch011

Mkrttchian, V., & Belyanina, L. (2016). The Pedagogical and Engineering Features of E- and Blended Learning of Adults Using Triple H-Avatar in Russian Federation. In V. Mkrttchian, A. Bershadsky, A. Bozhday, M. Kataev, & S. Kataev (Eds.), *Handbook of Research on Estimation and Control Techniques in E-Learning Systems* (pp. 61–77). Hershey, PA, USA: IGI Global; doi:10.4018/978-1-4666-9489-7.ch006

Mkrttchian, V., Bershadsky, A., Bozhday, A., Noskova, T., & Miminova, S. (2016). Development of a Global Policy of All-Pervading E-Learning, Based on Transparency, Strategy, and Model of Cyber Triple H-Avatar. In G. Eby, T. V. Yuser, & S. Atay (Eds.), *Developing Successful Strategies for Global Policies and Cyber Transparency in E-Learning* (pp. 207–221). Hershey, PA, USA: IGI Global; doi:10.4018/978-1-4666-8844-5.ch013

Mkrttchian, V., Bershadsky, A., Finogeev, A., Berezin, A., & Potapova, I. (2017). Digital Model of Bench-Marking for Development of Competitive Advantage. In P. Isaias & L. Carvalho (Eds.), *User Innovation and the Entrepreneurship Phenomenon in the Digital Economy* (pp. 288–303). Hershey, PA, USA: IGI Global; doi:10.4018/978-1-5225-2826-5.ch014

Mkrttchian, V., Kataev, M., Hwang, W., Bedi, S., & Fedotova, A. (2014). Using Plug-Avatars "hhh" Technology Education as Service-Oriented Virtual Learning Environment in Sliding Mode. In G. Eby & T. Vokan Yuzer (Eds.), *Emerging Priorities and Trends in Distance Education: Communication, Pedagogy, and Technology*. Hershey, PA, USA: IGI Global; doi:10.4018/978-1-4666-5162-3.ch004

Mkrttchian, V., Kataev, M., Hwang, W., Bedi, S., & Fedotova, A. (2016), Using Plug-Avatars "hhh" Technology Education as Service-Oriented Virtual Learning Environment in Sliding Mode. Leadership and Personnel Management: Concepts, Methodologies, Tools, and Applications (4 Volumes), (pp.890-902), IRMA, Hershey: PA, USA: IGI Global. Doi:10.4018/978-1-4666-9624-2.ch039

Mkrttchian, V., & Stephanova, G. (2013). Training of Avatar Moderator in Sliding Mode Control. In G. Eby & T. Vokan Yuzer (Eds.), *Project Management Approaches for Online Learning Design* (pp. 175–203). Hershey, PA, USA: IGI Global; doi:10.4018/978-1-4666-2830-4.ch009

Tolstykh, T., Vasin, S., Gamidullaeva, L., & Mkrttchian, V. (2017). The Control of Continuing Education Based on the Digital Economy. In P. Isaias & L. Carvalho (Eds.), *User Innovation and the Entrepreneurship Phenomenon in the Digital Economy* (pp. 153–171). Hershey, PA, USA: IGI Global; doi:10.4018/978-1-5225-2826-5.ch008

KEY TERMS AND DEFINITIONS

Avatar-Based Management: Is a control method and technique introduced by Vardan Mkrttchian in 2018.

Cyber Security Professionals: Are professionals of information security, within the framework of which the processes of formation, functioning and evolution of cyber objects are studied, to identify sources of cyber-danger formed while determining their characteristics, as well as their classification and formation of regulatory documents, implementation of security systems in future.

Intelligent and Analytics Platforms: Is engineering section of avatar-based management.

Machine Learning: Is a class of methods of artificial/natural intelligence, the characteristic feature of which is not a direct solution of the problem, but training in the process of applying solutions to a set of similar problems.

Optimizing Threat: Is situations at cyber security.

Self-Organizing Algorithm: Is section of avatar-based management use for cyber security.

Triangular Scheme of Machine Learning: Is engineering section of Machine learning techniques.

Variability Model: Is section of avatar-based management use for cyber security.

Chapter 6
Machine Learning With Avatar–Based Management of Sleptsov Net–Processor Platform to Improve Cyber Security

Vardan Mkrttchian
https://orcid.org/0000-0003-4871-5956
HHH University, Australia

Leyla Gamidullaeva
https://orcid.org/0000-0003-3042-7550
Penza State University, Russia

Sergey Kanarev
Penza State University, Russia

ABSTRACT

The literature review of known sources forming the theoretical basis of calculations on Sleptsova networks and on the basis of authors' developments in machine learning with avatar-based management established the basis for the future solutions to hyper-computations to support cyber security applications. The chapter established that the petri net performed exponentially slower and is a special case of the Sleptsov network. The universal network of Sleptsov is a prototype of the Sleptsov network processor. The authors conclude that machine learning with avatar-based management at the platform of the Sleptsov net-processor is the future solution for cyber security applications in Russia.

DOI: 10.4018/978-1-5225-8100-0.ch006

INTRODUCTION

Information and Communication Technology is acknowledged as crucial part of our current society, accessing within every level of our social environment. Along with the implementation of Information and Communication Technology, comes an important part of securing it. Their evolution and development has brought many benefits and have also given rise to cybercrime actors, serious cyber-attacks that had been demonstrated over the past few years. Cyber security has become an important subject of national, international, economic, and societal importance that affects multiple nations (Walker, 2012). Many countries have come to understand that this is an issue and has developed policies to handle this in an effort to mitigate the threats (Dawson, Omar, & Abramson, 2015). To address the issue of cyber security, various frameworks and models have been developed. Traditional approaches to managing security breaches is proving to be less effective as the growth of security breaches are growing in volume, variation and velocity (Bhatti & Sami, 2015). The purpose of this chapter is to show what future cyber security as engineering science and technology expects. In addition, the authors propose future solutions for the use of computer with a Sleptsov net processor when it will be actually created and practically implemented. The authors of the chapter did not consider the credibility issues of Sleptsov nets computing but completely trusted the creator of Sleptsov net as a processor, based on open sources, in particular on publications and webinars of IGI Global (Zaitsev, 2016; Zaitsev, et al., 2016; Zaitsev, 2018). Based solely on these publications in recent years in IGI Global and own experience, the authors research the emerging trends and perspectives of digital transformation of the economy using machine learning with avatar-based management at the platform of Sleptsov net processor and propose further prospects for development of hyper-computation.

BACKGROUND

Many researchers compare machine learning solutions for cyber security by considering one specific application (e.g., Buczak and Guven, 2016; Blanzieri and Bryl, 2008; Gardiner and Nagaraja, 2016) and are typically oriented to Artificial Intelligence experts.

The term "cyber security" refers to three things:

1. A set of activities and other measures, technical and non-technical, intended to protect computers, computer networks, related hardware devices and software, and the information they contain and communicate, including software and

data, as well as other elements of cyberspace, from all threats, including threats to national security;

2. The degree of protection resulting from the application of these activities and measures;

3. The associated field of professional endeavor, including research and analysis, aimed at implementing those activities and improving their quality (Jenab, et al., 2018).

At the same time, our previous research of the problem of cyber security showed that cyber security is a section of information security, within the framework of which the processes of formation, functioning and evolution of cyber objects are studied. It is necessary to identify sources of cyber-danger formed while determining their characteristics, as well as their classification and formation of regulatory documents, implementation of security systems in future. However, working on the application of the machine learning for Cyber Security applications with the use of developed by the authors Avatars-Based Management techniques, we came to the conclusion that this is not so, and the built-in cyber security systems can be destroyed by the same artificial intelligence.

The search for a solution to this discrepancy leads to a thought about advantages of natural intelligence displayed by humans where everything is interconnected, logical and protected (Mkrttchian, et al., 2015).

This paper is specifically aims to research the emerging trends and perspectives of digital transformation of the economy using machine learning with avatar-based management at the platform of Sleptsov net processor, and to identify their main limitations.

Sleptsov net concept mends the flaw of Petri nets, consisting in incremental character of computations, which makes Sleptsov net computing a prospective approach for ultra-performance concurrent computing (Zaitsev, 2018).

A Sleptsov net (SN) is a bipartite directed multi-graph supplied with a dynamic process (Zaitsev, 2016). An SN is denoted as $N=(P,T,W,\mu0)$, where P and T are disjoint sets of vertices called places and transitions respectively, the mapping F specifies arcs between vertices, and $\mu0$ represents the initial state (marking). The mapping $W: (P \times T) \rightarrow N \cup \{-1\}$, $(T \times P) \rightarrow N$ defines arcs, their types and multiplicities, where a zero value corresponds to the arc absence, a positive value – to the regular arc with indicated multiplicity, and a minus unit – to the inhibitor arc which checks a place on zero marking. N denotes the set of natural numbers. To avoid nested indices we denote w,i j j– = w(p, t) and+ =) . The mapping $\mu: P \rightarrow N$ specifies the place marking (Zaitsev, 2018).

Based on the previous research, performed by D. Zaitsev (2018; 2019), the main conclusion was drawn that Sleptsov networks are executed exponentially faster than Petri nets that makes it possible to recommend them as a parallel computing model for subsequent practical implementation.

Calculations on the networks of Sleptsov acquire all new applications presented in the works. First of all, computations on Sleptsov networks may be used for those applications in which parallel programming style can bring significant acceleration of computations.

Effective practical implementation of computations on Sleptsov networks requires the development of appropriate specialized automation systems for programming and hardware implementation of processors of Sleptsov networks. In addition, further development of theoretical methods of proving the correctness of programs in the language of Sleptsov networks and the development of universal networks that use mass parallelism are needed.

The advantages of computing on the Sleptsov networks are visual graphic language, the preservation of the natural domain parallelism, fine granulation of parallel computing, formal verification methods for parallel programs, fast mass-parallel architectures that implement the computation model (Zaitsev, 2018).

MAIN FOCUS OF THE CHAPTER

Issues, Controversies, Problems, Solutions and Recommendations

Machine learning (ML) was introduced in the late 1950's as a technique for artificial intelligence (AI) (Ambika, 2018). Over time, its focus evolved and shifted more to algorithms that are computationally viable and robust. One of the classical definitions of Machine Learning is the development of computer models for learning processes that provide solutions to the problem of knowledge acquisition and enhance the performance of developed systems (Duffy, 1995).

Machine learning is the use of artificial intelligence (AI) that provides systems with the capability to learn and automatically improve from experience (data) without being explicitly programmed. Machine learning focuses on developing computer programs that can access data and use them to learn by themselves.

Ideology of Machine Learning based on the principles of multiple use (reusability) and free distribution (share ability) copyright courses. Therefore, developers of training courses must adhere to generally accepted standards. To date, the most widely used models of the following courses:

1.   Model IEEE LOM (Learning Object Model), developed by the LTSC (Learning Technology Standard Committee) in 2002. The entire set of learning objects is divided into 9 composite hierarchies (categories): General (General), life cycle (Life Cycle), metadata (Metadata), technical (Technical), education (Education), legal (Rights), communication (Relation), annotation and classification (Annotation and Classification).

2.   System specifications consortium IMS (such as egg Content Packaging Specification, Metadata Specification, Digital Repositories Interoperability, Digital Repositories).

3.   Specifications Committee AICC (Aviation Industry Computer-Based Training Committee) originally intended for the development of computer-based training systems and technologies in the aviation industry.

4.   Specification SCORM (Shareable Course Object Reference Model), developed in the framework of the ADL (Advanced Distributed Learning), carried out by the Ministry of Defense. This is the industry standard for exchange of training materials based on tailored specifications ADL, IEEE, IMS. AICC. SCORM is the basis of the model modular design of educational material by separating the individual autonomous educational units (SCO - Shareable Content Objects) and their representation in the Web- specific repositories. SCO modules can be assembled together in various combinations and compiled into electronic textbooks using LMS- system. Thus, if in the first E-Learning systems, a teacher was expected to collect their own training courses to keep in his personal computer, and then manually organize a nationwide educational content, with the advent of such specifications as SCORM, this work is automated with the possibility of using Web 2.0 technologies and service-oriented approach.

Another pressing problem of modern machine learning systems is the problem of creating a student model based on the tracking of personal information related to learning trajectories passing on various training modules or web-services, the courses tendered tests. For these purposes, there is also a range of specifications, the most famous of which are:

- IEEE PAPI (Personal and Private Information);
- IMS LIP (Learner Information Package).

It uses the language XML (eXtension Markup Language) to write to the user's profile his curriculum vitae, teaching history, language skills, preferences to use computer platforms, passwords, access to training, etc. These data are then used to account for the individual characteristics of the student in determining the best means and methods of teaching.

Competency assessments for learning (learning competency assessment) are also used standardized specifications.

Having considerable experience in Intellectual Control and Communication, Avatar–Based Learning, Teaching and Training, and Avatar-Based Management, there is a need in modelling a joint system of Machine Learning with Avatar-Based Management with the use of Sleptsov net processor.

We define as an Avatar-Based Management (A-BM) model variability hypergraph VMG, consisting of two sets and predicate:

$$VMG = (V, U, P), \tag{1}$$

Set V describes the structure of a hypergraph on the vertex level:

$$V = \{v_{i,(weight)}\}, i = 1, 2, \ldots, N, \tag{2}$$

where N – is the total number of peaks is corresponding to the total number of characteristics of the A-BM model variability; *weight* - vertex weight in form n.1.1.1

Figure 1. Screen-stop of the modeling supercomputer with Sleptsov net processor to visualize of a report on the work of Cyber Security system

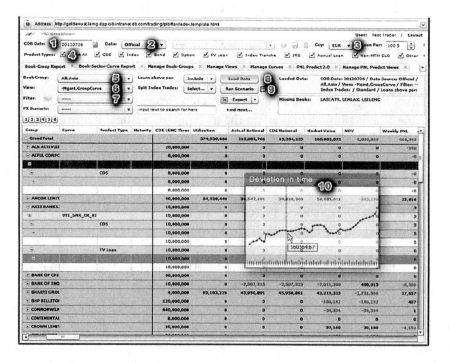

Figure 2. Block diagram of the model of platform for supercomputer with Sleptsov net processor in support cyber security applications

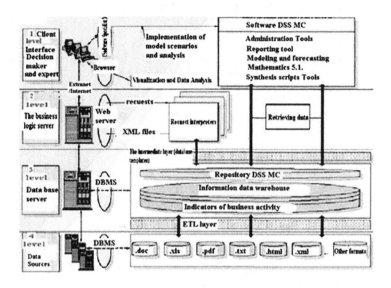

(Figure 3) corresponding to the index of the corresponding characteristics in the hierarchical structure of the A-BM model variability.

Set U has a capacity corresponding to the number of possible configurations of the A-BM:

$$U = \{u_j\} ; j = 1, 2, …, K, \tag{3}$$

where K - the numbers of hyperedges.

Obviously, depending on the size and structure of each elements of Block diagram the Model of Platform of the supercomputer with Sleptsov net processor in the Support Cyber Security Applications (fig.2), from the technical and communications capabilities available to the user at some point and some other features of the cardinality of the set U can vary significantly.

Predicates P - determines incidence of vertices and hyperedges of each layer. P is defined on the set of all pairs ($v \in V, u \in U$). Truth domain predicate P is the set R of variable cardinality $Bt \neq const$:

$$F(P) = \{(v, u) \mid P(v, u)_r\}, \tag{4}$$

Where $v \in V, u \in U, r \in R = \{1, 2… B_t\}$

Figure 3. Hypergaph, in which weights of vertices correspond to indices characteristics in the A-BM model supercomputer with Sleptsov net processor model and the hyper-cores correspond possible A-BM model configurations

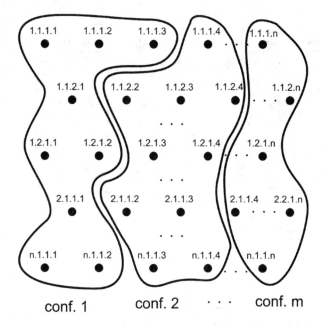

Variability of the cardinality of *R* is due to the same causes as the variability of U in Equation 3.

Considered a set-theoretic representation of the A-BM model to determine the variability of the matrix representation of this A-BM model is useful for creating software for Machine learning application with Avatar-Based Management technique use. Matrix representation (incidence matrix size *NxK*) hypergraph will have the form (Mkrttchian, et al., 2014) (5):

$$M_f = \| m_{ij} \|_{N \, x \, K},$$

(5)

where:

1, if $(v_i, u_j) \in F(P)$, $v \in V$, $u \in U$

$m_{ij} = 0$, if $(v_i, u_j) \notin F(P)$, $v \in V$, $u \in U$

In some cases it is more convenient to use the matrix of connected vertices of the hypergraph (Equation 6), which has a size *NxN* and reflects pairwise connectivity relations through vertices incident hyperedges.) .

$$M_c = \| m_{ij} \|_{N \times N},$$ (6)

where:

1, if (v_i, v_j) $\exists u_k$, $(v_i, u_k) \in F(P)$, $(v_j, u_k) \in F(P)$, $v \in V$, $u \in U$

$m_{ij} = 0$, *if (v_i, v_j) $\neg(\exists u_k)$, $(v_i, u_k) \in F(P)$, $(v_j, u_k) \in F(P)$, $v \in V$, $u \in U$*

Figure 4 shows the interaction model variability and configurations A-BM model depending on the success of the Machine learning applications, available technical and telecommunications capacity psychophysical characteristics of the people work in Cyber Security Systems. Based on the analysis of the current profile of this people, selects the optimal configuration of the A-BM (hyperedge in graph *VMG*), defined technical, interface and content component (vertices that is incident to this hyperedges).

Figure 4. Block diagram of the model of platform for supercomputer with Sleptsov Net processor

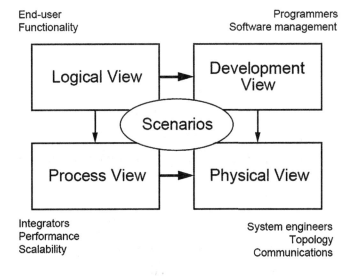

CONCLUSION

In this chapter, the authors studied machine-learning application with the use of Avatar-Based Management technique for cyber security issues. Our previous research devoted to the cyber security problem showed that cyber security is a section of information security, within the framework of which the processes of formation, functioning and evolution of cyber objects are studied. It is necessary to identify sources of cyber-danger formed while determining their characteristics, as well as their classification and formation of regulatory documents, implementation of security systems in future. However, working on the application of the machine learning for Cyber Security applications with the use of developed by the authors Avatars-Based Management techniques, we came to the conclusion that this is not so, and the built-in cyber security systems can be destroyed by the same artificial intelligence.

We proposed some prospects on Machine Learning with Avatar-Based Management at the Platform of the Sleptsov net processor, which are the Future Solutions to hyper-computations in the Support Cyber Security Applications.

ACKNOWLEDGMENT

The reported study was funded by RFBR according to the research project No. 18-010-00204_a.

REFERENCES

Ambika, P. (2018). Machine Learning. In P. Raj & A. Raman (Eds.), *Handbook of Research on Cloud and Fog Computing Infrastructures for Data Science* (pp. 209–230). Hershey, PA: IGI Global. doi:10.4018/978-1-5225-5972-6.ch011

Bhatti, B. M., & Sami, N. (2015). Building adaptive defense against cybercrimes using real-time data mining. *2015 First International Conference on Anti-Cybercrime (ICACC)*. 10.1109/Anti-Cybercrime.2015.7351949

Blanzieri, E., & Bryl, A. (2008). A survey of learning-based techniques of email spam filtering. *Artificial Intelligence Review*, *29*(1), 63–92. doi:10.100710462-009-9109-6

Buczak, A. L., & Guven, E. (2016). A Survey of Data Mining and Machine Learning Methods for Cyber Security Intrusion Detection. *IEEE Communications Surveys and Tutorials*, *18*(2), 1153–1176. doi:10.1109/COMST.2015.2494502

Dawson, M., Omar, M., & Abramson, J. (n.d.). Understanding the Methods behind Cyber Terrorism. *Encyclopedia of Information Science and Technology, 3*, 1539–1549. doi:10.4018/978-1-4666-5888-2.ch147

Duffy, J. (1975). IFToMM symposium—Dublin, September 1974. *Mechanism and Machine Theory, 10*(2-3), 269. doi:10.1016/0094-114X(75)90030-0

Gardiner, J., & Nagaraja, S. (2016). On the Security of Machine Learning in Malware C8C Detection. *ACM Computing Surveys, 49*(3), 1–39. doi:10.1145/3003816

Jenab, K., Khoury, S., & LaFevor, K. (2016). Flow-Graph and Markovian Methods for Cyber Security Analysis. *International Journal of Enterprise Information Systems, 12*(1), 59–84. doi:10.4018/IJEIS.2016010104

Jenab, K., Khoury, S., & LaFevor, K. (2018). Flow-Graph and Markovian Methods for Cyber Security Analysis. In I. Management Association (Ed.), Cyber Security and Threats: Concepts, Methodologies, Tools, and Applications (pp. 674-702). Hershey, PA: IGI Global. doi:10.4018/978-1-5225-5634-3.ch036

Mkrttchian, V. (2015). Use Online Multi-Cloud Platform Lab with Intellectual Agents: Avatars for Study of Knowledge Visualization & Probability Theory in Bioinformatics. International Journal of Knowledge Discovery in Bioinformatics, 5(1), 11-23. Doi:10.4018/IJKDB.2015010102

Mkrttchian, V., & Aleshina, E. (2017). *Sliding Mode in Intellectual Control and Communication: Emerging Research and Opportunities*. Hershey, PA: IGI Global. doi:10.4018/978-1-5225-2292-8

Mkrttchian, V., & Belyanina, L. (Eds.). (2018). *Handbook of Research on Students' Research Competence in Modern Educational Contexts*. Hershey, PA: IGI Global. doi:10.4018/978-1-5225-3485-3

Mkrttchian, V., Bershadsky, A., Bozhday, A., & Fionova, L. (2015). Model in SM of DEE Based on Service-Oriented Interactions at Dynamic Software Product Lines. In G. Kurubacak & T. Yuzer (Eds.), *Identification, Evaluation, and Perceptions of Distance Education Experts* (pp. 231–248). Hershey, PA: IGI Global. doi:10.4018/978-1-4666-8119-4.ch014

Mkrttchian, V., Bershadsky, A., Bozhday, A., Kataev, M., & Kataev, S. (Eds.). (2016). *Handbook of Research on Estimation and Control Techniques in E-Learning systems*. Hershey, PA: IGI Global. doi:10.4018/978-1-4666-9489-7

Mkrttchian, V., Kataev, M., Shih, T., Kumar, M., & Fedotova, A. (2014). Avatars "HHH" Technology Education Cloud Platform on Sliding Mode Based Plug-Ontology as a Gateway to Improvement of Feedback Control Online Society. International Journal of Information Communication Technologies and Human Development, 6(3), 13-31. Doi:10.4018/ijicthd.2014070102

Mkrttchian, V., Veretekhina, S., Gavrilova, O., Ioffe, A., Markosyan, S., & Chernyshenko, S. V. (2019). The Cross-Cultural Analysis of Australia and Russia: Cultures, Small Businesses, and Crossing the Barriers. In U. Benna (Ed.), *Industrial and Urban Growth Policies at the Sub-National, National, and Global Levels* (pp. 229–249). Hershey, PA: IGI Global. doi:10.4018/978-1-5225-7625-9.ch012

Walker, S. (2012). Economics and the cyber challenge. *Information Security Technical Report*, *17*(1-2), 9–18. doi:10.1016/j.istr.2011.12.003

Zaitsev, D. A. (2018). Sleptsov Net Computing. In M. Khosrow-Pour (Ed.), Encyclopedia of Information Science and Technology (4th ed.; pp. 7731-7743). Hershey, PA: IGI Global. doi:10.4018/978-1-5225-2255-3.ch672

Zaitsev, D. A. (2019). Sleptsov Net Computing. In M. Khosrow-Pour (Ed.), Advanced Methodologies and Technologies in Network Architecture, Mobile Computing, and Data Analytics (pp. 1660-1674). Hershey, PA: IGI Global. doi:10.4018/978-1-5225-7598-6.ch122

ADDITIONAL READING

Bershadsky, A., Bozhday, A., Evseeva, J., Gudkov, A., & Mkrtchian, V. (2017). Techniques for Adaptive Graphics Applications Synthesis Based on Variability Modeling Technology and Graph Theory, In A. Kravets, M. Shcherbakov, M. Kultsova, & O. Shabalina, (Eds.), *Proceedings of CIT&DS 2017*, (pp. 169–179). Switzerland: Springer International Publishing AG. DOI: 10.1007/978-3-319-65551-2_33

Bershadsky, A., Evseeva, J., Bozhday, A., Gudkov, A., & Mkrtchian, V. (2015), Variability modeling in the automated system for authoring intelligent adaptive applications on the basis of three-dimensional graphics, In A. Kravets, M. Shcherbakov, M. Kultsova, & O. Shabalina, (Eds.), *Proceedings of CIT&DS 2015*, (pp. 149–159). Switzerland: Springer International Publishing AG. DOI: 10.1007/978-3-319-23766-4

Glotova, T., Deev, M., Krevskiy, I., Matukin, S., Mkrttchian, V., & Sheremeteva, E. (2015). Individualized learning trajectories using distance education technologies, In A. Kravets, M. Shcherbakov, M. Kultsova, & O. Shabalina, (Eds.), *Proceedings of CIT&DS 2015*, (pp. 778–793). Switzerland: Springer International Publishing AG. DOI: 10.1007/978-3-319-23766-4

Mkrttchian, V. (2011). Use 'hhh" technology in transformative models of online education. In G. Kurubacak & T. Vokan Yuzer (Eds.), *Handbook of research on transformative online education and liberation: Models for social equality* (pp. 340–351). Hershey, PA, USA: IGI Global; doi:10.4018/978-1-60960-046-4.ch018

Mkrttchian, V. (2012). Avatar manager and student reflective conversations as the base for describing meta-communication model. In G. Kurubacak, T. Vokan Yuzer, & U. Demiray (Eds.), *Meta-communication for reflective online conversations: Models for distance education* (pp. 340–351). Hershey, PA, USA: IGI Global; doi:10.4018/978-1-61350-071-2.ch005

Mkrttchian, V. (2015). Modeling using of Triple H-Avatar Technology in online Multi-Cloud Platform Lab. In M. Khosrow-Pour (Ed.), Encyclopedia of Information Science and Technology (3rd Ed.). (pp. 4162-4170). IRMA, Hershey: PA, USA: IGI Global. Doi:10.4018/978-1-4666-5888-2.ch409

Mkrttchian, V. (2016). The Control of Didactics of Online Training of Teachers in HHH University and Cooperation with the Ministry of Diaspora of Armenia. In V. Mkrttchian, A. Bershadsky, A. Bozhday, M. Kataev, & S. Kataev (Eds.), *Handbook of Research on Estimation and Control Techniques in E-learning systems* (pp. 311–322). Hershey, PA, USA: IGI Global; doi:10.4018/978-1-4666-9489-7.ch021

Mkrttchian, V., & Aleshina, E. (2017). The Sliding Mode Technique and Technology (SM T&T) According to Vardan Mkrttchian in Intellectual Control(IC). In *Sliding Mode in Intellectual Control and Communication: Emerging Research and Opportunities* (pp. 1–9). Hershey, PA: IGI Global; doi:10.4018/978-1-5225-2292-8.ch001

Mkrttchian, V., Amirov, D., & Belyanina, L. (2017). Optimizing an Online Learning Course Using Automatic Curating in Sliding Mode. In N. Ostashewski, J. Howell, & M. Cleveland-Innes (Eds.), *Optimizing K-12 Education through Online and Blended Learning* (pp. 213–224). Hershey, PA, USA: IGI Global; doi:10.4018/978-1-5225-0507-5.ch011

Mkrttchian, V., Aysmontas, B., Uddin, M., Andreev, A., & Vorovchenko, N. (2015). The Academic views from Moscow Universities of the Cyber U-Learning on the Future of Distance Education at Russia and Ukraine. In G. Eby & T. Vokan Yuzer (Eds.), *Identification, Evaluation, and Perceptions of Distance Education Experts* (pp. 32–45). Hershey, PA, USA: IGI Global; doi:10.4018/978-1-4666-8119-4.ch003

Mkrttchian, V., & Belyanina, L. (2016). The Pedagogical and Engineering Features of E- and Blended Learning of Adults Using Triple H-Avatar in Russian Federation. In V. Mkrttchian, A. Bershadsky, A. Bozhday, M. Kataev, & S. Kataev (Eds.), *Handbook of Research on Estimation and Control Techniques in E-Learning Systems* (pp. 61–77). Hershey, PA, USA: IGI Global; doi:10.4018/978-1-4666-9489-7.ch006

Mkrttchian, V., Bershadsky, A., Bozhday, A., Noskova, T., & Miminova, S. (2016). Development of a Global Policy of All-Pervading E-Learning, Based on Transparency, Strategy, and Model of Cyber Triple H-Avatar. In G. Eby, T. V. Yuser, & S. Atay (Eds.), *Developing Successful Strategies for Global Policies and Cyber Transparency in E-Learning* (pp. 207–221). Hershey, PA, USA: IGI Global; doi:10.4018/978-1-4666-8844-5.ch013

Mkrttchian, V., Kataev, M., Hwang, W., Bedi, S., & Fedotova, A. (2014). Using Plug-Avatars "hhh" Technology Education as Service-Oriented Virtual Learning Environment in Sliding Mode. In G. Eby & T. Vokan Yuzer (Eds.), *Emerging Priorities and Trends in Distance Education: Communication, Pedagogy, and Technology*. Hershey, PA, USA: IGI Global; doi:10.4018/978-1-4666-5162-3.ch004

Mkrttchian, V., Kataev, M., Hwang, W., Bedi, S., & Fedotova, A. (2016), Using Plug-Avatars "hhh" Technology Education as Service-Oriented Virtual Learning Environment in Sliding Mode. Leadership and Personnel Management: Concepts, Methodologies, Tools, and Applications (4 Volumes), (pp.890-902), IRMA, Hershey: PA, USA: IGI Global. Doi:10.4018/978-1-4666-9624-2.ch039

Mkrttchian, V., & Potapova, I. (2018). Professors of Innovative Implementations in Sliding Mode Digital Technology for Enhancing Students Competence. In Mkrttchian, V., & Belyanina, L., (Eds.) Handbook of Research on Students' Research Competence in Modern Educational Contexts, (189-204), Hershey, PA, USA: IGI Global. Doi:10.4018/978-1-5225-3485-3.ch010

Mkrttchian, V., & Stephanova, G. (2013). Training of Avatar Moderator in Sliding Mode Control. In G. Eby & T. Vokan Yuzer (Eds.), *Project Management Approaches for Online Learning Design* (pp. 175–203). Hershey, PA, USA: IGI Global; doi:10.4018/978-1-4666-2830-4.ch009

Mkrttchian, V., Vasin, S., Surovitskaya, G., & Gamidullaeva, L. (2018). Improving the mechanisms of formation of MS students' research competencies in Russian core universities. In Mkrttchian, V., & Belyanina, L., (Eds.), Handbook of Research on Students' Research Competence in Modern Educational Contexts, (90-105), Hershey, PA, USA: IGI Global. Doi:10.4018/978-1-5225-3485-3.ch005

Tolstykh, T., Vasin, S., Gamidullaeva, L., & Mkrttchian, V. (2017). The Control of Continuing Education Based on the Digital Economy. In P. Isaias & L. Carvalho (Eds.), *User Innovation and the Entrepreneurship Phenomenon in the Digital Economy* (pp. 153–171). Hershey, PA, USA: IGI Global; doi:10.4018/978-1-5225-2826-5.ch008

Tolstykh, T., Vertakova, J., & Shkarupeta, E. (2018). Professional Training for Structural Economic Transformations Based on Competence approach in the Digital Age. In Mkrttchian, V., & Belyanina, L., (Eds.) Handbook of Research on Students' Research Competence in Modern Educational Contexts, (209-229), Hershey, PA, USA: IGI Global. Doi:10.4018/978-1-5225-3485-3.ch011

KEY TERMS AND DEFINITIONS

Hypercomputation or Super-Turing Computation: Is a multi-disciplinary research area with relevance across a wide variety of fields, including computer science, philosophy, physics, electronics, biology, and artificial intelligence; models of computation that can provide outputs that are not Turing computable.

Machine Learning: Is the use of artificial intelligence (AI) that provides systems with the capability to learn and automatically improve from experience (data) without being explicitly programmed.

Machine Learning Application With Avatar-Based Management Technique Use: Is a class of methods of natural intelligence, the characteristic feature of which is not a direct solution of the problem, but training in the process of applying solutions to a set of similar problems.

Shareable Content Object Reference Model (SCORM): Is a collection of standards and specifications for web-based electronic educational technology (also called e-learning).

Sleptsov Net (SN): Is a bipartite directed multi-graph supplied with a dynamic process.

Chapter 7
Intelligent Log Analysis Using Machine and Deep Learning

Steven Yen

https://orcid.org/0000-0001-8378-0687
San Jose State University, USA

Melody Moh

https://orcid.org/0000-0002-8313-6645
San Jose State University, USA

ABSTRACT

Computers generate a large volume of logs recording various events of interest. These logs are a rich source of information and can be analyzed to extract various insights about the system. However, due to its overwhelmingly large volume, logs are often mismanaged and not utilized effectively. The goal of this chapter is to help researchers and industrial professionals make more informed decisions about their logging solutions. It first lays the foundation by describing log sources and format. Then it describes all the components involved in logging. The remainder of the chapter provides a survey of different log analysis techniques and their applications, consisting of conventional techniques using rules and event correlators that can detect known issues, plus more advanced techniques such as statistical, machine learning, and deep learning techniques that can also detect unknown issues. The chapter concludes describing the underlying concepts of the techniques, their application to log analysis, and their comparative effectiveness.

DOI: 10.4018/978-1-5225-8100-0.ch007

INTRODUCTION

Long before the advent of computers, logging has been used in various fields. Examples included physical logbooks, accounting transaction ledgers, car maintenance records, etc. They are used to record any events of interest based on the context. The information in the logs can then be used in the future for troubleshooting purposes, help improve operating procedures, act as an audit trail, and so on.

The practice of logging was adopted in computing systems from the very beginning. Developers used printf statements throughout their code to print relevant information to help them debug the code when issues arise. Some of the messages are only used during development and are removed before release, others were placed strategically to help with troubleshooting or monitoring purposes later on. These log messages can be shown directly to the user or be sent to specific outputs channels such as to a file. Due to its usefulness, logging became common practice, and nowadays almost every piece of software has logging capability. In modern computing systems, logs can come from operating systems, network devices, and various application software. They are meant to record interesting events that occurred when programs are ran.

These logs from various devices and processes proved to be extremely useful for the detection of security issues. Operating system logs (or *host logs*) can be analyzed to detect unauthorized access, such as that by an attacker using a *ssh-scanner* (Chuvakin, Schmidt, & Philips, 2013). Network logs can be analyzed to detect unusual traffic such as that between a malware and a remote attacker's device (Stamp, 2006). Web application logs can be analyzed to detect attacks such as cross-site scripting, SQL injection, and invalid resource access (Liang, Zhao, & Ye, 2017). Many, if not all, cyberattacks leave traces in logs somewhere, one just needs to know what to look for.

However, because of the automated nature of log generation in computing systems, the volume of logs generated became very large. An unfortunate consequence of this is that many users began to view logs as an annoyance rather than a helpful tool. Logs were seldom looked at and are often simply deleted when space runs out. To address these issues, log management systems were developed to facilitate the collection, storage, and analysis of logs.

Log analysis can be done manually by inspecting raw text files directly or using event viewers provided by log management systems. Such manual inspection is labor-intensive and often not timely enough for real-time incident response. To address these limitation, rule-based systems were developed that can evaluate log events based on a library of known issues (known as a rule-base). These tools proved to be quite effective and have helped organizations prevent many incidents in a timely fashion. The drawback is that they can only detect known issues for which there are exact rules in the rule-base, and misses unknown issues. To help detect new and unknown issues, anomaly detection approaches were introduced, which are based on identifying unusual or abnormal behavior. Statistical, machine learning, and deep learning techniques proved to be quite suitable for this application, because they can form their own detection criteria from training data rather than relying on human operators to specify rules. Over the years, more and more of these techniques have been applied to log analysis with impressive results.

Cognitive science played an important role in the development of intelligent log analysis tools. This is because the individuals using the computer systems, the analysts, and the attackers are all human. Understanding the thought process and motives of all these individuals is therefore crucial in trying to identify issues and malicious activity. As testament to this, there has been studies done to understand how human subjects analyze logs to detect issues (Layman, Diffo, & Zazworka, 2014) as well as studies to understand user web browsing behavior through logs (Kussul & Skakun, 2005). In fact, intelligent log analysis tools have always been designed to emulate human, with early rule-based systems aiming to capture the decision making process of security experts who wrote the rules. However, the attack methods are constantly evolving as the creators of these attacks are humans too, whose goals are to come up with new ways to avoid detection. It takes time for security experts to study and dissect a new attack, write a good rule, and add the rule to the rule-base. Machine Learning (ML) and Deep Learning (DL) techniques, which aim to emulate humans' ability to understand data and identify patterns, were soon applied to log analysis for the detection of issues. The authors will present ML techniques that emulate humans' ability to organized and group data based on similarity, as well as DL techniques that further aims to emulate the biological structure of the human brain (i.e. Artificial Neural Networks).

In this chapter, the authors first provide an overview of log management and analysis, describing important concepts and common practices. Then, the chapter dives in and focuses on different log analysis techniques, starting with common and conventional techniques and build towards more advanced and novel techniques.

BACKGROUND

Source, Content, and Format of Logs

To facilitate troubleshooting and monitoring of computer systems, developers have devised logging mechanisms to record various events of interest. Software at all levels can generate logs, from the operating systems and drivers to user applications. In addition to hosts and servers, logs can also be generated by network devices such as firewalls, switches, routers, etc. Any network enabled devices such as printers can also generate logs.

At minimum, these log messages contain 3 components:

- **Timestamp:** Exact time and date at which an event occurred
- **Source:** The device that generated the message
- **Data:** Description of the event

The data portion of the log message is meant to provide detailed information about the event needed for troubleshooting or monitoring. It can contain additional fields such as message type, priority, severity, and other application-specific categorization. Below figures show examples of logs generated by 3 different systems.

Figure 1. Example Apache Log Messages for HTTP Request Events

```
eventslb@localhost:/var/log/apache2$ sudo tail -3 access.log
::1 - - [11/Dec/2018:15:25:16 -0800] "GET /testwebsite/AirlinesHome.html HTTP/1.1"
 200 3161 "-" "Mozilla/5.0 (X11; Ubuntu; Linux i686; rv:60.0) Gecko/20100101 Firef
ox/60.0"
::1 - - [11/Dec/2018:15:25:50 -0800] "GET /testwebsite/airlinelist.html HTTP/1.1"
404 518 "http://localhost/testwebsite/AirlinesHome.html" "Mozilla/5.0 (X11; Ubuntu
; Linux i686; rv:60.0) Gecko/20100101 Firefox/60.0"
::1 - - [11/Dec/2018:15:26:05 -0800] "POST /testwebsite/AirlinesHome.html HTTP/1.1
" 200 3161 "http://localhost/testwebsite/AirlinesHome.html" "Mozilla/5.0 (X11; Ubu
ntu; Linux i686; rv:60.0) Gecko/20100101 Firefox/60.0"
eventslb@localhost:/var/log/apache2$
```

Figure 2. Example Log Messages by a Linux Operating System

```
eventslb@localhost:/var/log$ sudo head -5 /var/log/syslog
Dec 10 21:56:42 localhost rsyslogd: [origin software="rsyslogd" swVersion="8.16.0" x-pid="835
gd was HUPed
Dec 10 21:56:42 localhost sudo: pam ecryptfs: pam_sm_authenticate: /home/eventslb is already
Dec 10 21:56:43 localhost anacron[768]: Job `cron.daily' terminated
Dec 10 21:56:43 localhost anacron[768]: Normal exit (1 job run)
Dec 10 21:57:06 localhost com.canonical.indicator.application[1238]: (process:1549): indicato
tion already exists, re-requesting properties.
eventslb@localhost:/var/log$
```

Figure 3. Example Log Messages by a Network Intrusion Detection System (NIDS) Snort

```
12/01-20:33:45.678230  [**] [1:10000002:1] ICMP test detected!! [**] [Classification: Generic IC
MP event] [Priority: 3] {ICMP} 192.168.80.11 -> 192.168.8.20
12/01-20:33:46.679906  [**] [1:10000002:1] ICMP test detected!! [**] [Classification: Generic IC
MP event] [Priority: 3] {ICMP} 192.168.80.11 -> 192.168.8.20
12/01-20:33:47.682380  [**] [1:10000002:1] ICMP test detected!! [**] [Classification: Generic IC
MP event] [Priority: 3] {ICMP} 192.168.80.11 -> 192.168.8.20
12/01-20:33:48.685613  [**] [1:10000002:1] ICMP test detected!! [**] [Classification: Generic IC
MP event] [Priority: 3] {ICMP} 192.168.80.11 -> 192.168.8.20
```

As can be seen, the formatting of log messages can differ significantly depending on the source.

Some standards have been created in an attempt to standardize logs. Some examples include W3C Extended Log File Format, Cisco Security Device Event Exchange (SDEE), Syslog Format, etc. Unfortunately, adoption of such standardized format is not widespread. Most applications still use non-standard formats and therefore have to be treated as free-form text by system administrators (Chuvakin et al., 2013).

As for the file type, the logs can be directly human-readable such as ASCII or Unicode text, or XML marked-up documents. Logs can also be stored as binary files that cannot be read by humans directly, and require special programs (compatible log reading tools) to make them readable by humans. As an example, the Windows Event Log files are stored in binary format and need to be read using the Windows Event Viewer (Chuvakin et al., 2013).

Binary log formats cannot be inspected directly using text editors or searched with utilities. However, log files stored in binary formats have the advantage of taking less space (for the same amount of information) and can be written and accessed more efficiently by the specialized programs. Furthermore, binary logs often allow compression and encryption.

Yet another way to store log data is in a relational database, which can also be considered a binary format. Long-term storage of log data is usually in a relational database. This also allows querying of the data with a more universal interface that is not product specific.

Log Management

Modern computing systems and applications generate large volumes of logs. These logs can also come from many different sources. Proper log management is needed to make good use of these logs.

Log management is performed by software that is usually part of the operating system. An example of such a system is the Windows Event Viewer. Windows Event Viewer has been a mainstay of Windows operating systems since Windows NT. It provides an interface for OS users or system administrators to manage the logs generated by various processes. Users can configure various aspects such as the size-limit of log files, what to do when log files become full (overwrite, archive, or stop logging), and define actions to be completed when certain events arise. Unix systems have similar mechanisms to manage the log files in the /var/log directory.

Due to the distributed nature of modern computing systems, logs from multiple hosts and devices on a network often need to be collected in a centralized location to allow efficient storage and analysis. This is achieved by the use of *loghosts*, which are dedicated machines on the network to which logs are sent. The hosts and devices in the network are configured to send their logs to the *loghost* over the network using a protocol such as Syslog, SNMP, or FTPS.

The most popular protocol used to send logs over a network is the Syslog protocol (note: in addition to being a protocol, Syslog also defines some formatting as discussed previously). Syslog is supported by various Unix systems. Syslog is also compatible with Windows Servers with the use of commercial applications that converts log entries from the Windows Event Log format into Syslog format before forwarding the entries to the Syslog server. Syslog can be implemented on top of the User Datagram Protocol (UDP) or the Transmission Control Protocol (TCP) depending on the application and whether receipt confirmation is required. Syslog follows a client-server model, where the *loghost* acts as the server and the hosts and devices generating logs act as the clients. The *loghost* then saves the logs to its disk storage, optionally in a relational database (Chuvakin et al., 2013).

Simple Network Management Protocol (SNMP) is a protocol used for transporting logs that is commonly used with networking devices like switches and routers. It uses trapping and polling mechanisms to transport information. When something of interest happens at a device, it sends a trap request along with the log message to a Network Management Station (NMS), which behaves like a *loghost*. The NMS also uses a polling mechanism to query devices for predefined variables and statuses (Chuvakin et al., 2013).

As an alternative of having *loghosts* on an organization's own network, an organization can also subscribe to log management services in the cloud, such as those offered by Splunk. The user simply configures the devices on their network to send their logs to the provider's server. Users can then customize and view their logs through a web interface.

Log Analysis

The purpose of log analysis is to derive meaning and useful information from log data. Log analysis is often done at the centralized location where the logs are sent, such as the *loghost*. However, some log analysis can also be performed locally at a device, such as by a firewall. The benefit of performing log analyses at a centralized location is that one can determine correlations between different components and derive global insight.

Due to the largely unstandardized nature of log formats from different sources, it is necessary to convert logs into a common format. This process is known as *normalization*. We define the free-form raw text as *log message*, and the normalized form of each message as a *log event*. In this process, each log message is parsed, or scanned from beginning to end, to extract relevant pieces of information and place them in the appropriate fields of the corresponding log event. There can be standard as well as user-defined fields that are deemed useful for the purpose of searching and filtering. Some common fields used are: timestamp, priority, protocol, source IP, etc. If a field is not applicable to a certain log message, then it can be set to NULL or left empty. It is also common to have a field called "raw log" that stores the original unaltered log message, the purpose of which is to preserve any information that was not extracted but might be useful in the future (Chuvakin et al., 2013).

Once all the log messages are converted into log events of a common format, the data can be filtered easily based on specific fields, various statistics can be calculated, and aggregate analyses can be performed. Additionally, this opens the door to more complex log analysis techniques such as those using machine learning and deep learning as we'll discuss in the following chapters.

RULE-BASED LOG ANALYSIS

The earliest form of log analysis techniques to be applied were rule-based. Users define rules that describe the type of events or sequences of events to look for, which often correspond to known issues. This type of analysis is also often referred to as *event correlation*, because we're analyzing disparate events (from different time and sources) to try to determine if some situation of interest is taking place. For example, to detect potential attacks, a system administrator might define a rule to flag all the login-related events between 12:00 AM and 6:00 AM (assuming the company has no operations in other time-zones).

The most primitive way to perform rule-based analysis is by searching through the log files or, in the case where logs were normalized and stored in databases, by writing complex queries that represent the rule. Unfortunately, this manual approach is neither efficient nor timely, and issues may not be detected until long after they have occurred. As a result, modern computing systems often use a software known as a rule engine (or event correlator) to perform this kind of rule-based analysis automatically. Rule engines monitor events in real-time and checks them against a list of rules in a rule-set (which is analogous to the virus definitions of anti-virus software). The rules usually describe the pattern of events to look for and the actions to take when there's a match. An example of such a rule engine is the Simple Event Correlator (SEC), which is an open-source tool that can be used to analyze log events. SEC can analyze events from operating systems, firewalls, intrusion detection systems (IDS), etc., against a rule-set to identify issues and take appropriate actions (Chuvakin et al., 2013).

To correlate events over time, the event correlators need to be able to remember past events. While past events are persisted on disk, accessing disk is time-consuming and inefficient; therefore, most event correlators store the information needed for analysis in the memory. These systems can be categorized into the following 2 types based on how they store historical information in the memory (Chuvakin et al., 2013):

1.   Stateful Rule Engine
2.   Stream-Based Engine

Stateful Rule Engine

A Stateful Rule Engine maintains state information for each rule. Event patterns are represented as a sequence of state transitions (Chuvakin et al., 2013). For example, we may be interested in detecting *unauthorized* port-scans (port-scan in itself doesn't necessarily mean an attack; it could be done by an internal device to detect available services). To detect unauthorized port-scans, we can use the following rule (written as a pseudo-code):

- **Pattern:** An event E1 from the IDS with type = "port-scan" is followed by an event E2 from the firewall with type = "packet-rejected" where the source and destination IP addresses of E1 and E2 are the same.
- **Action:** Alert system administrator by sending the message "potential unauthorized port-scan detected!"

To implement the pattern, the rule engine could define the following 2 states. Where the detection of an event of type "port-scan", will cause the transition from state A to B.

- **State A:** This is the initial state where no port-scan had occurred. When in this state, we're monitoring for *any* events of type "port-scan" from the IDS logs.
- **State B:** A port-scan event was observed in the past with some source and destination IP address pair (src_ip_i, dst_ip_i). When in this state, we're monitoring for events of type "packet-rejected" from the firewall with the same IP address pair (src_ip_i, dst_ip_i).

We could have multiple instances of this rule, one for each IP address pair (src_ip_i, dst_ip_i). In State B, if an event of type "packet-rejected" comes in with the same IP address pair, the action defined in the rule (notify sys admin) would be triggered.

As we've seen, there could be many instances of this rule for different IP address pairs. Additionally, an instance of the rule may stay in State B indefinitely (i.e. in the case of an *authorized* port-scan). As a result, the rule engine could exhaust the memory trying to maintain all these state information. For this reason, it is imperative that the rule engine have an eviction policy to discard instances of the rule from memory. The most common way to do this is to give each rule instance a time-to-live (TTL), after which it is evicted. For example, in the case of an actual unauthorized port-scan, the "port-scan" event from the IDS and the "packet-rejected" event from the firewall should happen fairly close to each other. So we can set a TTL of say 3 minutes for each rule instance in state B. If no "packet-rejected" event arrives within 3 minutes, it is likely that the firewall allowed the packet to pass through, which means the port-scan was from an authorized host. Aside from such rule-specific TTL, most Stateful Rule Engines have a default TTL value (Chuvakin et al., 2013).

Stream-Based Engine

Stream-Based Engine, also known as Complex Event Processing (CEP) engine, enforce rules in a different way. Instead of storing instances of rules in different states waiting to be satisfied (or evicted), Stream-Based Engine maintains queues of recent events in the memory and apply the rules to the events in this queue. As new events are added to the queue, old events are evicted (in the order they came in). In other words, the events are "streamed" through the queue (Chuvakin et al., 2013).

The memory usage of Stream-Based Engine is controlled by the length of the queue. However, this queue needs to be sized appropriately to be able to detect certain events. If new events are coming in really fast (high throughput), the queue will need to be very long, requiring significant memory.

The choice of the type of event correlator (Stateful Rule Engine or Stream-Based Engine) depends on the application. Stateful Rule Engines are very versatile, and new rules can be added easily without requiring other configuration changes. On the other hand, new rules for Stream-Based Engine may require the queue length be adjusted. In fact, even a change in event throughput may require the queue lengths be adjusted. The strength of Stream-Based Engine is that it can be much faster than Stateful Rule Engines. It is not uncommon for systems to incorporate both types of event correlators.

ANOMALY-BASED APPROACH FOR LOG ANALYSIS

The rule-based correlation techniques for log analysis discussed in the last section is efficient for detecting known attacks or issues for which we can write a rule. However, rule-based techniques falls short for new or unknown attacks and issues. This is where anomaly-based techniques come in.

Anomaly-based techniques aim to detect behavior that is abnormal or unusual. However, to determine if something is abnormal, we need to first establish what is normal. This is where baselining comes in. That is, we use historical data from a time period during which the behavior is believed to be normal to train a statistical or machine learning model such that the model recognizes what is normal. Then, the model can be applied to new events to determine if they are significantly different from what is considered normal (Stamp, 2006).

In this section, we'll discuss different techniques used for anomaly detection.

Statistical Techniques

We consider statistical techniques for anomaly-detection to be any techniques that use statistical concepts. This includes techniques that uses statistic measures to describe data distribution, as well as machine learning techniques that uses statistics in conjunction with specialized algorithms to allow the model to learn.

Distribution Statistics

Behavior of computer systems can be approximated by normal distributions. That is, there are some normal, expected behaviors that occur with the highest likelihood, and some behaviors different from the norm that occur with lower likelihood. The more some behavior differs from what is normal, the lower the likelihood of its occurrence. Distribution statistics allow us to quantify the likelihood and identify anomalies.

To characterize the behavior of computer systems, we derive specific features (i.e. measurements) from the log events. Common features of interest are the frequencies of different types of events. Such features can be easily derived from normalized log messages based on the timestamp and type fields. Example features include the number of user login attempts per hour, the number of HTTP requests with 500 status code per hour, and so on. In our anomaly-detection system, the values for each of these features could be modeled as a normal distribution, and the distribution statistics would be used to establish acceptable bounds for the feature.

For a given feature, the analysis involves the following steps (Chuvakin et al., 2013):

1. Determine a past window or time period over which to establish the baseline (e.g. the last month).
2. Calculate the mean (μ) and standard deviation (σ) of the feature's value over this time period with the following formulas (where x_i is the i-th value of the feature, and N is the total number of values observed in the baseline period):

$$\mu = \left(\sum_{i=1}^{N} x_i \right) / N \tag{1}$$

$$\sigma = \sqrt{\frac{\sum_{i=1}^{N} \left(x_i - \mu \right)^2}{N}} \tag{2}$$

Calculate the standard error (SE) with the formula:

$$SE = \frac{\sigma}{\sqrt{N}} \tag{3}$$

Establish an allowable interval by adding and subtracting $1.96*SE$ from the mean. This will act as the baseline. The multiplier 1.96 is recommended by Chuvakin et al. (2013), but other values can be used depending on the desired accuracy.

3. Compare future data points to the allowable interval, and flag them as anomalies if they fall outside of the interval.
4. Periodically reevaluate the baseline with new data to account for normal (legitimate) change in system behavior.

The above analysis is effective in detecting issues that can manifest in single features. For example, a brute-force attack on a password could be detected by performing this analysis on the feature *login attempts per hour*. However, not all issues are like this. To detect some issues require considering multiple features together, such as in the machine learning techniques we'll discuss next.

Machine Learning Techniques

Machine learning techniques allow a computer program to learn from some sample (training) data, and makes predictions about future data. They often leverage statistical techniques along with specialized algorithms to achieve this.

One major advantage of machine learning over distribution statistics analysis is that they can consider multiple features at the same time to obtain a more global view of the system. To process multiple features simultaneously, we will form feature vectors where each element is a feature. For example, a feature vector at time t might take on the following form:

$$x_t = \begin{bmatrix} number\ of\ logins\ per\ hour_t \\ number\ of\ logoffs\ per\ hour_t \\ no.\ of\ failed\ logins\ per\ hour_t \\ no.\ of\ events\ to\ port\ 22\ per\ hour_t \\ \vdots \end{bmatrix} \tag{4}$$

The process of creating these feature vectors is known as feature extraction, and is required by most machine learning techniques. Once the training data is converted to this format, we can begin training the model.

Machine learning algorithms can be classified into to 2 main categories: supervised versus unsupervised. Supervised techniques are those for which the training data is labeled with ground-truth. In the context of log analysis, this would mean that, for each training sample (feature vector), we know the corresponding class label, such as what type of traffic it is (Brute-Force Attack, SYN Flood, normal, etc.). Alternatively, it may also be sufficient to just label traffic as normal vs abnormal. Examples of this type of techniques include k-Nearest Neighbor (kNN), Logistic Regression, Support Vector Machines, just to name a few. These types of techniques are useful for known problems, and can complement rule-based systems by removing the burden of writing rules (Chuvakin et al., 2013). However, labeling large volumes of training data could be labor-intensive or even infeasible.

Unsupervised techniques are those for which the training data does not need to be labeled. The algorithms attempt to discover some underlying structure of the dataset on their own. Examples of unsupervised techniques include k-Means Clustering, Principle Component Analysis (PCA), and others. These techniques are advantageous in that they remove the need to label data. Note, in the context of anomaly detection, a technique can be considered unsupervised if it is only trained on the normal data points during the training phase. The labels aren't considered by the algorithm. In the research studies, the abnormal data points are set aside for the sole purpose of measuring the models performance during the detection phase.

In recent studies, it had become common to quantify the accuracy of machine learning algorithms using the statistics *Precision, Recall*, and *F-Measure*, given by the following equations:

$$Precision = \frac{TP}{TP + FP} \tag{5}$$

$$Recall = \frac{TP}{TP + FN} \tag{6}$$

$$F\,Measure = \frac{2 \cdot Precision \cdot Recall}{Precision + Recall} \tag{7}$$

where *TP, FP, FN* represents the number of *True Positives, False Positives*, and *False Negatives*, respectively (Du et al., 2017).

In the context of anomaly detection, the *Precision* quantifies what portion of the points identified by the model as anomalous are actually anomalous (per the label). The *Recall* quantifies what portion of the points that are actually anomalous the model successfully detected. The *F-measure*, given as a harmonic mean of *Precision* and *Recall* represents a holistic measure that takes into account both (Du et al., 2017). For all three measures, the higher the better.

k-Means Clustering

Clustering is the grouping of similar data points. That is, data points will be divided into groups (clusters), and members of each group will be similar to each other. There are many different clustering algorithms, of which k-Means is the most popular.

Given a set of data points $D = \{x_1, x_2, x_3, ...x_n\}$ and the desired number of clusters k, the k-Means algorithm will iteratively partition the data points based on some distance measure, until in the end there are k clusters (Liu, 2012).

In our application, each data point x_a will be a feature vector. If there are d features, then each data point will be a feature vector of d-dimensions. That is, $x_a = <x_{a1}, x_{a2}, x_{a3}, ..., x_{ai}, ..., x_{ad}>$, where x_{ai} denotes the *i-th* feature of the feature vector

$\mathbf{x}_a$. Then, to measure the distance of this data point to another point, say $\mathbf{x}_b$, we use the Euclidean norm, given by equation (8):

$$x_a - x_b = \sqrt{\sum_{i=1}^{d}\left(x_{ai} - x_{bi}\right)^2} \qquad (8)$$

Intuitively, the distance is inversely related to similarity. The smaller the distance between two feature vectors, the more similar they are to one another. The greater the distance between two feature vectors, the more different they are to one another.

The k-Means algorithm also uses the notion of a centroid, which is the center (or mean) of a cluster of points over all dimensions. Given a set of data points in the same cluster, say C_q, the centroid, $\mathbf{m}_q$, of the cluster is calculated by taking the mean of each dimension across all data points. That is, the *i-th* element of the centroid (m_{qi}) is given by equation (9):

$$m_{qi} = \frac{\sum_{x \in C_q} x_i}{|C_q|} \qquad (9)$$

With these notions defined, we can describe the k-Means algorithm, which consist of the following steps:

1. Randomly select k data points from the training set D to be the initial, dummy centroids.
2. For each data point in D, compute with equation (8) the distance from that data point to each of the k centroids. Assign the data point to the same cluster as the closest centroid.
3. For each of the k clusters that results from step 2 above, calculate the new centroid for the cluster with equation (9).
4. Repeat steps 2 and 3 until the stopping criteria is met. The cluster assignments for the data points are finalized.

The stopping criteria to use is at the discretion of the user. Common stopping criterion used can be (1) when no data point changes cluster membership anymore, (2) when the centroid of clusters stop changing, or (3) when the sum-of-squared-error (SSE) from all points to the centroid of the cluster the point belongs to drops below some threshold (Liu, 2011).

In this algorithm, it is up to the user to specify the value of k, which determines the number of resulting clusters. One can simply run the algorithm with different k values and compare the resulting clusters analytically, or pick k from recommended ranges based on literature for the application (Han, Kamber, & Pei, 2012).

Researchers proposed the following method to use k-Means for anomaly detection. First, the k-Means algorithm is ran on the historical data from a period of normal operation, yielding clusters of different types of normal traffic. These clusters and their centroids will then represent our baseline, normal behaviors. To evaluate new events, we extract the feature vector, then calculate the distance between it and the closest centroid. If this distance is above some threshold λ, then we've detected an anomaly. Different ways of selecting this threshold has been explored in the papers by Lima et al. (2010) and Yin et al. (2015). The selection of this threshold has direct impact on the accuracy and the rate of false positives.

Lima et al. (2010) and Yin et al. (2015) both implemented anomaly detection systems based on k-Means and applied them to the dataset from the 1999 International Knowledge Discovery and Data Mining Tools Competition (KDD Cup), which is a competition organized by the Association for Computing Machinery (ACM). The dataset, widely referred to as KDD Cup 1999, contains network traffic data over a nine-week period. The data was preprocessed to have various features vectors extracted. Additionally, each feature vector is labeled with the traffic it represents, from the classes *normal*, *port sweep*, *back*, *smurf* and many other malicious traffic types. In anomaly detection, the goal is to identify abnormal data points, and there's no need to classify the exact type of abnormal traffic. Therefore, to evaluate their anomaly detection systems, Yin et al. (2015) extracted 2000 data points with the label *normal* and 100 abnormal data points. The 100 abnormal data points were drawn from data points with the labels *teardrop*, *back*, *smurf*, and *neptune*. The accuracy of their anomaly detection system based on traditional k-Means is summarized in Table 1.

Table 1. Anomaly Detection Results from Applying Traditional k-Means to KDD CUP 1999

k (Number of Clusters)	Average Detection Rate	Average False Positive Rate
5	44.4%	0
6	58.6%	0.37%
7	64%	0.34%
8	64%	0.36%
9	93.2%	0.54%
10	78.8%	0.24%

(Yin et al., 2015)

These authors did not use the standard metrics *precision*, *recall*, and *F-Measure*. Instead, they defined *detection rate*, which is the rate at which anomalous points were correctly detected (similar to *recall*), and *false positive rate*, which is the rate at which normal points were incorrectly flagged as anomalous.

The goal is to maximize the *detection rate* and to minimize the *false positive rate*. Unfortunately, these goals are often competing, where a high *detection rate* often leads to a high *false positive rate*. So, coming up with a good compromise based on the system need is crucial. In Table 1, we see that with k=9, the k-Means model achieved a fairly high *detection rate* of 93.2% and a low *false positive rate* of 0.54%, which is a decent combination for the performance.

As shown in these experiments, the k-Means algorithm produces fairly good accuracy when used for anomaly detection. Another advantage of k-Means is that the algorithm is very intuitive and easy to implement. Finally, it is also relatively efficient compared to other algorithms, with a run time of $O(n*k*t)$, where n is the number of data points in the training set, k is the targeted number of clusters, and t is the number of iterations (Liu, 2011).

The weakness of the algorithm is that it is sensitive to noise and outliers, which can skew final cluster centroids significantly (Han, Kamber, & Pei, 2012), so the training data must be selected carefully or be screened properly. Additional challenges of k-Means are that the accuracy is affected by model parameters such as the number of clusters (k) and the distance threshold (λ), which need to be adjusted when the algorithm is applied to different datasets. Even with fixed parameters, the selection of initial centers (which is typically done randomly) also affects the accuracy (Yin et al., 2015). The need for customization results in additional work for the users.

Principle Component Analysis (PCA)

PCA is an unsupervised machine learning technique that leverages linear algebra concepts. The goal of PCA is to reduce the dimensionality of the dataset. For example, if our feature vectors have n dimensions (i.e., n features), the goal would be to reduce the dimension of the vectors to less than n without significant loss of information.

Again, we perform feature extraction on the baseline log data to get a set of feature vectors $D = \{x_1, x_2, x_3, ..., x_i, ..., x_n\}$, where x_i represents the i-th feature vector.

Then, the steps of the PCA algorithm are as follows (Han, Kamber, & Pei, 2012):

1. Normalize the feature vectors such that the value range of each element is the same (e.g., from 0 to 1).
2. Compute the principle components of the data points. The principle components are n orthogonal unit vectors that will act as the new axes for the data set.

Conceptually, the principle components are identified by finding new axes along which the data varies the most.

3. Linearly transform the data points to represent them as coordinates along the principle components (the new axes).

4. Sort the principle components based on variance of all the data points along it. The more the data points vary along an axis, the more important or influential the corresponding principle component is. In other words, we're sorting the principle components (axes) from most influential to least influential.

5. The axes with the least influence can be disregarded or dropped completely.

Figure 4 from the interactive website http://setosa.io/ev/principal-component-analysis/ illustrates this process very well (Powell).

In the example shown in Figure 4, we start with a data set with 2 features, x and y. Each data point can be represented by a 2D vector, where the first element is the value along the x-axis, and the second element is the value along the y-axis. Figure 4(a) shows the data points plotted along x-y axes. Using PCA, we compute 2 orthogonal unit vectors pc1 and pc2 as shown by the bolded lines in Figure 4(a). Through linear transformation, we change the representation of the data points. In the vector associated with each data point in the new representation, the first element is the value along pc1, and the second element is the value along pc2; that is, we've changed the axes. Figure 4(b) shows the data points plotted along the new axes. As can be observed in Figure 4(b), the data points varies a lot more along pc1 than along pc2, indicating that pc1 has more influence on the data than pc2. To reduce the dimension of the data set, we can drop the axis (pc2) that has little influence.

Figure 4. PCA Example with 2D Data Set

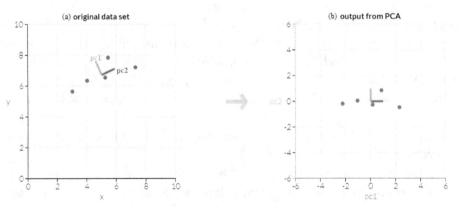

Source: generated on interactive PCA introduction webstie http://setosa.io/ev/principal-component-analysis/

The data set can be analyzed based on the remaining dimension (pc1) alone without significant loss of information.

According to Xu et al. (2009), the principle component along which there is high variation captures significant correlations between the original features. For instance, from Figure 4(a) above, we see that pc1 has a direction and slope that is similar to the trend produce by the majority of the data points (higher x results in higher y); that is, pc1 describes some underlying correlation between the original features x and y. For such correlation to emerge requires that significant number of data points follow this pattern, meaning it is a normal, expected behavior. On the other hand, principle components along which there are low data variation (such as pc2), indicates a correlation described by it is weak or absent—that is, the correlation is not exhibited by many of the data points. Researchers use this intuition as basis for PCA-based anomaly detection techniques (Xu et al., 2009).

In the PCA based anomaly-detection model proposed by Xu et al. (2009), principle components along which there are high data variation are referred to as normal subspaces (or normal directions), while principle components along which there are low data variation are referred to as abnormal subspaces (or abnormal directions).

Anomaly detection is done by looking for isolated points along the abnormal subspaces that are distant from the rest of the points. The intuition is that an isolated, distant point along the abnormal subspace is exhibiting a correlation that is not exhibited by the majority of points, so that point is likely to be anomalous. To evaluate a particular sequence of events, we first extract the feature vector (which represents a point in the multi-dimensional space), then we project the point on to

Figure 5. Simplified Example of PCA-Based Anomaly Detection

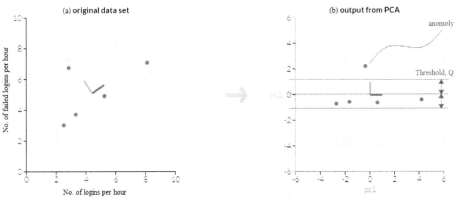

Source: generated on interactive PCA introduction website http://setosa.io/ev/principal-component-analysis/

the abnormal subspaces and calculate the length of the projection, and if it is above some threshold Q, it is deemed abnormal (Xu et al., 2009). More detail on this process and the selection of the threshold Q can be found in the paper by Xu et al. (2009). Figure 5 illustrates the usage of PCA for anomaly detection in a simplified 2D example.

As shown in the example in Figure 5, we have a data set with 2 features "No. of failed logins per hour" and "No. of logins per hour". The dataset plotted with the original features as axes is shown in Figure 5(a). Then, we perform PCA to identify the principle components pc1 and pc2, and transform the dataset. The transformed data set plotted with pc1 and pc2 as axes is shown in Figure 5(b). The principle component pc1 has the highest data variation, and is therefore our normal subspace. The principle component pc2 has very little data variation, and is therefore our abnormal subspace. We inspect data points along pc2 and flag as anomalous points at distances of more than Q away. In this example, just one point is flagged as an anomaly as shown in Figure 5(b).

Multiple researchers, including Xu et al. (2009), He et al. (2016), and Du et al. (2017), have applied PCA to detect anomalies from system logs. A common dataset used by all three projects is the dataset generated by Xu et al. (2009). This dataset consists of HDFS logs from running Hadoop map-reduce jobs on a cluster of 203 virtual machines in AWS for 48 hours. The dataset was subsequently labeled to identify normal versus abnormal traffic for benchmarking purposes. Table 2 below shows the experiment results obtained by He et al. (2016) comparing the performance of PCA and a clustering based technique that is very similar to k-Means based on the author's description.

As shown in Table 2, PCA have a higher precision than the clustering technique, but a lower recall. The overall F-Measures are almost the same. The choice of which algorithm to use is based on the application. For example, if a user wants to target close to 100% detection of anomalies and don't mind having to manually screen out some false alarms, then the user might prefer PCA over clustering in this application.

The advantage of PCA is that it requires less customization than other machine learning algorithms, as the values for the hyperparameters such as the Q threshold can be automatically determined by the system based on a formula (Xu et al., 2009).

Table 2. Anomaly Detection Results from Applying PCA and Clustering to HDFS Dataset

Algorithm	Precision	Recall	F-Measure
PCA	0.98	0.67	0.79
Clustering	0.87	0.74	0.80

(He et al., 2016)

In the process of reducing the dimensions, PCA also helps identify the relevant features to monitor. Finally, PCA is highly scalable, with a run time that is linear in the number of feature vectors (Xu et al., 2009).

A disadvantage of PCA is that it is less intuitive compared to other ML algorithms. The linear transformations underlying PCA are often treated as a black box by the operators (Xu et al., 2009). This means that when the model isn't performing as expected, it is challenging for the operator to determine what the issue is. Furthermore, PCA is sensitive to certain dataset characteristics, and the performance may be different when applied to different datasets (He et al., 2016).

Deep Learning Techniques

Artificial Neural Networks (ANNs) have been proposed as a computation model to solve problems as far back as 1943 (Russell & Norvig, 2010). Unfortunately, due to the limitation of computational power and limited volume of data, ANNs were regarded as impractical. Due to the improved computation capacity and the availability of large-scale datasets in recent decades, ANNs have re-gained popularity as tools to solve difficult problems, and showed impressive results.

The design of ANN is inspired by the animal brain, which consists of intricate networks of specialized cells known as neurons. Each neuron can detect electrochemical input signals upstream, and relay some signal downstream to other neurons if the linear combination of the input exceeds some threshold (i.e. the neuron activates or "fires"). The signal sent downstream by a neuron is referred to as its *output activation*. A network of these neurons work together to allow animals to perceive inputs from the environment, make decisions, and perform various tasks. Researchers designed ANN to mimic the brain's construction in hopes of creating a computational model that can solve diverse problems.

ANN consists a network of computational units known as nodes that mimic the neuron. A node have input links upstream that receive signals, and output links downstream for sending out signals. Each link have a weight associated with it that represents the strength of that connection (how much influence it should have). A node will compute the linear combination of the input signals and the weights, apply an activation function to calculate an output value (referred to as the *activation*), and relay it downstream through all the output links.

Nodes can be classified into three types. Hidden nodes (or hidden units) are those whose input links as well as output links connect to other nodes. Input nodes (or input units) are those whose input links is connected to the network input from the environment; together, all the input nodes form what is called the input layer. Output nodes (output units) are those whose output link is connected to the network output (which contributes to the result); together, all the output nodes form what is called the output layer (Russell & Norvig, 2010). Hidden nodes can be organized into different hidden layers based on how many upstream links they are from the input layer. A deep network is an ANN that involves one or more hidden layers. Deep Learning is the application of such deep networks to solve problems.

The neural network can be thought of as a function f that takes the features as network input and outputs a prediction. That is, $f(x) = \hat{y}$, where x is the input feature vector and $\hat{y}$ is the network's prediction. This function f is parameterized by the weights of the links connecting the nodes. Training a neural network boils down to using training samples to help update the weights so as to minimize the error between the network's output prediction $\hat{y}$ and what the correct (or ground-truth) value should be, y.

Since its introduction, numerous different types of ANN have been developed, including Multilayer Perceptron (MLP), Convolutional Neural Networks (CNN), Recurrent Neural Networks (RNN) and many more. For each of the ANN types, there are also multiple configurations and model hyperparameters that can be varied, such as the number of layers, the number of nodes per layer, and the type of connections involved. This results in infinite number of possible ANN architectures. The architecture choice should be based on the application.

Multilayer Perception (MLP)

Multilayer Perception (MLP) is one of the earliest and most basic ANN architecture proposed. MLP networks are feedforward and fully-connected. Feedforward means that information only flows in one direction, from the input layer through successive hidden layers then out through the output layer; output links of each layer can only connect to the input of downstream layers, and not to any upstream layer's input. Fully-connected means that each node is connected to all the nodes in the subsequent layer. Figure 6 shows an example of a MLP network.

The number of input nodes is the number of features (or attributes) each data point has. The number of output nodes depends on whether we're dealing with a classification problem or a regression problem, as well as how we're representing the output. If we're dealing with a classification problem, it is typical to have the same number of output nodes as there are categories. For classification problems, the output layers will also typically have the *softmax* function applied, which will turn the output into a probability distribution across a set of output categories. The category with the highest value corresponds to the output category predicted by the model.

MLP—and ANNs in general—can be represented as weight matrices. That is, we store all the link weights as elements in weight matrices. There would a separate

Figure 6. Example MLP Network

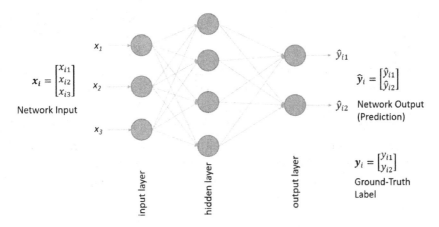

weight matrix for each layer. For example, the MLP in Figure 6 would have a 3x4 matrix W_1 storing the weights from the input to the hidden layer; then, it would also have a 4x2 matrix W_2 storing the weights from the hidden layer to the output layer. The input to the network would be represented as vectors where each element is a feature. Likewise, the network output would be represented as a vector with the same dimension as the number of output nodes.

During the training phase, each sample input (feature vector) x_i is fed in through the network input. Then, the neurons in each layer calculates the linear combination of the weights and activations of its input links, apply its activation function to determine its own output activation, and pass it to the next layer. These computations are performed by the nodes in each layer, one layer at a time, until we reach the output layer and get the network output vector, $\hat{y}_i$, which is the network's prediction. We call this entire process the *forward pass*, and it can be done efficiently by performing a series of matrix multiplications and element-wise application of the activation functions.

After the forward pass, the network's output vector $\hat{y}_i$ is compared with the ground-truth label y_i to calculate the error. The error gradient is then passed backwards through the network to update the weights in a process known as *backpropagation*, which again can be done efficiently as a series of matrix multiplications. See the book *Deep Learning* by Goodfellow et al. for a detailed discussion on training neural networks.

Using MLP, anomaly detection can be treated as a binary classification problem with two classes: normal and anomalous. Malaiya et al. (2018) implemented and evaluated an MLP (referred to as FCN, or Fully-Connected Network) whose outputs are passed in to a softmax function to yield a probability distribution across the two classes "Normal" and "Attack", and utilized the cross entropy loss function for backpropagation. This model was tested on a Honeypot dataset from Kyoto University as well as the NSL-KDD dataset. The results are summarized in Table 3.

The NSL-KDD dataset is a modified version of the KDD Cup 1999 dataset we discussed previously, with many redundant records removed. The Kyoto-Honeypot dataset contains network traffic between 2009 and 2018 collected in a honeypot system operated by Kyoto University (Malaiya et al., 2018).

In addition to FCN, Malaiya et al. (2018) experimented with conventional machine learning models as well as other deep learning models. Their experiments showed that deep learning models, including FCN, were much more accurate than conventional machine learning models. However, FCN was less accurate than RNN-based deep

Table 3. Anomaly Detection Results from Applying FCN to Kyoto-Honeypot and NSL-KDD Datasets

Dataset	Precision	Recall	F-Measure
Kyoto-Honeypot	0.99	0.87	0.93
NSL-KDD*	-	-	0.87

(Malaiya et al., 2018)

*The researchers did not list the precision and recall measures from applying FCN to the NSL-KDD dataset in their report.

Figure 7. Anomaly Detection Results of Different Deep Learning Models Applied to Kyoto-Honeypot

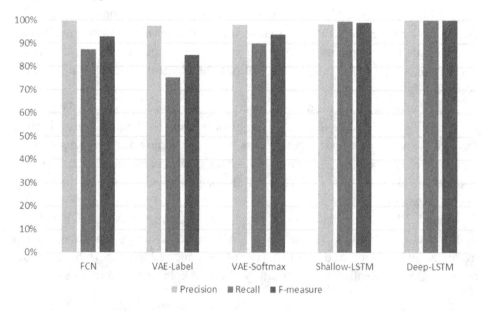

learning models like Long Short-Term Memory (LSTM), which we will discuss in the next section. Figure 7 shows the results of comparing different deep learning models with the Kyoto-Honeypot dataset (Malaiya et al., 2018).

The benefit of FCN is that it consumed the least time for training compared to all the other deep learning models considered by Malaiya et al. (2018). This is likely because the simple, feedforward construction of FCN makes it more parallelizable than other deep learning techniques.

A disadvantage of the FCN anomaly detection model devised by Malaiya et al. (2018) is that it requires both "normal" and "attack" samples in the training phase. In many applications, "attack" samples may not be available, or may make up a much smaller proportion of the dataset than "normal" samples. The Kyoto-Honeypot dataset is unique in that it is from a honeypot, which is a decoy system designed to bait or attract attackers, resulting in it seeing more malicious traffic than usual. Most other systems will see significantly more normal traffic than malicious traffic.

Another disadvantage of FCN is that, like conventional machine learning techniques, it requires the input to be feature vectors. As we'll see in the next section, RNN-based deep learning models can directly process sequences of events, without the need to extract feature vectors, resulting in less data preprocessing workload.

Recurrent Neural Networks (RNN)

Recurrent Neural Networks (RNN) is a class of ANN that involves recurrent connections. Recurrent connections are links that go from the output of a node back to its own input or to the input of a node upstream of it (in a previous layer). The existence of these recurrent connections is what distinguishes RNNs from feedforward networks like MLP. Figure 8 shows an example of a basic RNN, where each hidden node have output links that connect to its own input (a self-loop) as well as the inputs of other hidden nodes in the same layer.

The existence of recurrent connections allow RNNs to have memory, as outputs from a previous time step is relayed forward continually. So when the network is processing input from a certain time step, it considers both current information as well as information from the past (from the recurrent connections). This mechanism makes RNNs ideal for processing sequences.

To help illustrate how RNN processes sequences of inputs, the literature often shows RNN in an "unrolled" view as shown in Figure 9. In Figure 9(b), multiple networks are shown, but they all represent the same (single) network, just shown at different time steps.

The recurrent connections from the output of the hidden layer nodes back to their own inputs are shown in orange in Figure 9. The information passed along these

Figure 8. Basic RNN Example

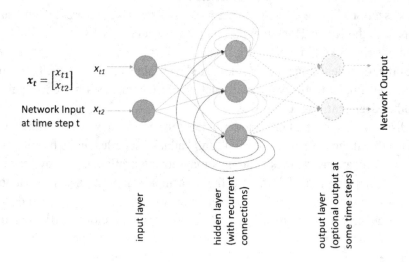

Figure 9. RNN Sequence Processing

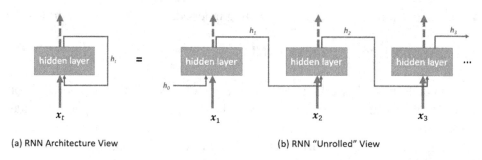

(a) RNN Architecture View   (b) RNN "Unrolled" View

recurrent connections can be thought of as an internal state maintained and updated by the network, and is often referred to as the hidden state, and can be represented as a vector h_t where each element is the output from one of the hidden nodes. Note, in the very first time step, the hidden state vector h_0, is usually initialized as a vector of 0s.

Note, in both Figure 8 and Figure 9, the connection from the hidden layer to the network output are shown as dashed lines. This is because, depending on the application, we may or may not need output at each time-step. Consider the problem of text sentiment analysis: our input sequence may be a sentence (such as a tweet), but we only require an output at the end that says whether the sentiment is "positive" or "negative". In this application, we can disregard the outputs from all the time steps except for the last one.

Due to its ability to process sequences, it did not take long before RNNs became widely adopted for various applications such speech recognition, translation, time-series data analysis, and not surprisingly, log analysis. To tackle various problems and meet special needs, many variations of RNN have been developed.

The basic RNN architecture shown above is susceptible to the *vanishing gradient problem* or the *exploding gradient problem* when applied to really long sequences. Vanishing gradient is due to the repeated multiplications of small numbers together that results in smaller and smaller products (i.e. vanishing). On the other hand, exploding gradient results from repeatedly multiplying very large numbers resulting in larger and larger products (i.e. exploding).

The backpropagation phase of training is where these problems can manifest, since as the error gradients are pass backwards through the network, repeated multiplications with the weights of the links occur. This problem is especially pronounced in RNNs, as the error gradient not only flow backwards layer by layer, but also flows backwards in time in a process known as Back-Propagation Through Time (BPTT), which results in even more multiplications (Hochreiter & Jurgen, 1997).

Various techniques were developed to address the vanishing or exploding gradient problems, one of which is the introduction of gates. Gates are structures that control the flow of information through links in the network. They can amplify or lessen the effect of specific links dynamically. The gates are placed strategically to remedy the vanishing and exploding gradient problems. Specific RNN architectures that uses gates include Gated Recurrent Units (GRUs) and Long Short-Term Memory (LSTM) networks.

Long Short-Term Memory (LSTM)

The Long Short-Term Memory (LSTM) architecture was first proposed by Hochreiter & Schmidhuber (1997). It was designed to remedy the vanishing gradient and exploding gradient problems of training regular RNNs over long sequences. It also allowed the network to have long term memories from many time-steps ago (Hochreiter & Schmidhuber, 1997).

LSTM introduced 3 gates, called the output gate, input gate, and forget gate. Figure 10 shows a simplified view of the LSTM with the gates. Together, the gates control the information flow, essentially allowing the network to selectively "forget" irrelevant information, and "remember" relevant information. By actively managing what is remembered or forgotten is what allows LSTM to have very long term memories. See the blog by Olah (2015) for a much more detailed description of LSTM.

The other function of these gates is that they allow the LSTM network to enforce constant error flow through certain units, preventing the error gradient from exploding or vanishing (Hochreiter & Schmidhuber, 1997).

Figure 10. Simplified View of LSTM

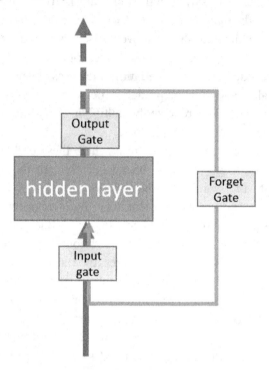

As log analysis essentially consists of processing a time-series of log events, where the events often have long-term dependencies, LSTM was applied to log analysis and showed much better accuracy than MLP.

Because of LSTM's ability to capture time-dependent relationships, the preparation of the input data can also be simplified. Instead of having to mine for features like "No. of Logins per Hour" to capture time-dependent relations, we can feed sequences of events directly to the LSTM network.

An approach adopted by many researchers is to parse the time-series log messages by extracting "log keys" (or "message types") and parameter values. Log keys represent the constant parts of the log message, while parameters represent the runtime values or identifiers that can change. For example, for the log message "XMT: Solicit on eth0, interval 128060ms", the number 128060 is a parameter value that can change, while the rest of the string stays constant. As such, the information we would extract from this log message could be a parameter value $v=128060$ and a log key $k =$ "XMT: Solicit on eth0, interval *ms", where we've put a placeholder "*" in place of the parameter. The next time a log message like this comes up, it will have the same log key, but a different parameter value (Du, Li, Zheng, & Srikumar, 2017). Because the log messages are generated by print statements in the code, we know there are a finite number of distinct log keys that can show up. We'll denote the set of all distinct log keys as $K=\{k_1, k_2, k_3, ..., k_n\}$, where there are n distinct log keys total.

Using this method, we parse the log messages one by one, recording the log key and the parameter values for each. Table 4 shows an example extracting the log key and parameter values from a few sample messages.

By parsing the logs this way allows us to construct a log key sequence, as well as one or more parameter value sequence(s) for each log key. For example from Table 4 we can obtain the log key sequence $[k_{23}, k_{16}, k_{23}, k_{23}, k_{16}, ...]$, a parameter value sequence for k_{23} of [128060, 124170, 109280, ...], and a parameter value sequence for k_{16} of ["192.168.56.1", "129.168.56.2", ...]. Additionally, temporal information can be preserved by recording the time-elapsed between messages as another parameter value for each log key. Each of these sequences can then be processed with a LSTM network to model their behavior. Most previous work used only the log key sequence for anomaly detection and showed satisfactory results. After a first pass analyzing just the log key sequence, specific parameter sequences can be analyzed if higher accuracy is desired (Du et al., 2017).

For this application, we use the LSTM in a many-to-one setup, where we'll have the LSTM ingest a sequences of a few inputs, and output a prediction of the next log key at the last time-step. To do this, we feed the LSTM output at the last time-step into one or more fully-connected (FC) layers, the output of which is then sent through a softmax function to produce a probability distribution across all the possible log keys. The key with the highest probability will be the network's prediction for the next log key. That is, we've built a classifier that predicts the next log key based on the last few log keys. Figure 11 shows an example setup.

Table 4. Sample Log Key and Parameter Extraction

Timestamp	Message	Log Key	Parameter Value
11/10/18 3:45:42	XMT: Solicit on eth0, interval 128060ms	k_{23} = "XMT: Solicit on eth0, interval *ms"	128060
11/10/18 4:08:45	refused connect from 192.168.56.1	k_{16} = "refused to connect from *"	192.168.56.1
11/10/18 4:19:29	XMT: Solicit on eth0, interval 124170ms	k_{23} = "XMT: Solicit on eth0, interval *ms"	124170
11/10/18 4:34:01	XMT: Solicit on eth0, interval 109280ms	k_{23} = "XMT: Solicit on eth0, interval *ms"	109280
11/10/18 4:44:44	refused connect from 192.168.56.2	k_{16} = "refused to connect from *"	129.168.56.2
...	...	...	...

Figure 11. Log Key Processing with LSTM Network

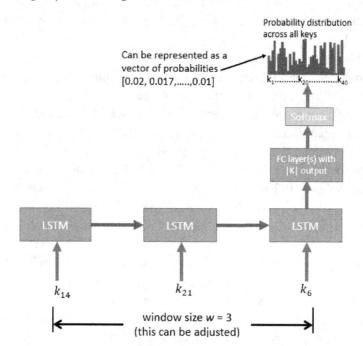

The number of inputs the LSTM ingests before its output is passed into the FC layers is a hyperparameter that can be tuned. This is shown as the window size (w) in Figure 11.

We need to format our training data appropriately to train this network. From the log key sequence, we need to construct input-label pairs (training samples) where the input consists of w consecutive log keys, and the output consists of the next log key in the sequence. For example, if our log key sequence from the training data set consists of the sequence $[k_{14}, k_{21}, k_6, k_{28}, k_4, k_{35}]$, then we can derive the following 3 input-label pairs (training samples):

- $[k_{14}, k_{21}, k_6] \rightarrow k_{28}$
- $[k_{21}, k_6, k_{28}] \rightarrow k_4$
- $[k_6, k_{28}, k_4] \rightarrow k_{35}$

For each training sample, the input is fed into the network, which outputs a probability distribution across the |K| categories (all possible log keys). This is then compared to the 1-hot encoding of the correct label to calculate the error, which is then back-propagated through the network to update and train the weights in the network. We train the network on all the training samples from our baseline period, then we're ready to use the model for anomaly detection.

In the detection (prediction) phase, we feed the most recent w log keys into the network to get the output probability distribution across the all the possible log keys, then we sort these log keys base on probability from highest to lowest and pick the top g of these as the candidates of the next log key predicted by our network. If the log key of the next event that arrives is one of the top g candidates predicted by our network, then the event is deemed normal, otherwise it is deemed anomalous. The value of g is another hyperparameter that can be tuned by the user. The researchers have tried different values of g, and the results showed that g of around 5 was suitable for the datasets tested (Du et al., 2017).

Additionally, the LSTM units can be layered to form deeper models, where the output from one LSTM layer goes into the input of another LSTM layer rather than to the FC output layer directly. Both Du et al. (2017) and Malaiya et al. (2018) studied the effects of the number of layers. In Figure 7, the *Shallow-LSTM* model has one LSTM layer, whereas the *Deep-LSTM* model had three LSTM layers (Malaiya et al., 2018). In general, the deeper the model (the more layers), the higher the accuracy, but it also requires more training time compared to shallower models (Malaiya et al., 2018). For most purposes, one layer is sufficient.

Du et al. (2017) tested their LSTM-based model (which they referred to as "DeepLog") as well as a conventional PCA model on two different datasets, with the results shown in Figure 12. One dataset they used is generated from an experimental OpenStack system deployed on CloudLab, performing VM-related tasks. This dataset was highly imbalanced, with about 94% of "normal" data points and 6% of "anomalous" data points. The other dataset they used is the HDFS dataset generated by Xu et al. (2009) that we discussed previously; this dataset was more balanced, with about 75% of "normal" data points and 25% of "abnormal" data points.

As can be seen in Figure 12, the LSTM-based model (DeepLog) was more accurate than the PCA model on both datasets, with higher F-Measures. Additionally, the LSTM model's accuracy was fairly consistent across the two datasets, indicating that LSTM models are less sensitive to specific characteristics of datasets (Du et al., 2017). This implies that one can apply LSTM models on different types of logs and achieve similar accuracy without significant customization.

Based on the experiments by Du et al. (2017) and Malaiya et al. (2018), we can see that RNNs, specifically LSTMs, are very effective for log analysis, and is more accurate than all conventional machine learning techniques as well as other deep learning techniques.

Figure 12. Anomaly Detection Results Comparing LSTM-Based Model (DeepLog) and PCA on HDFS and OpenStack Datasets

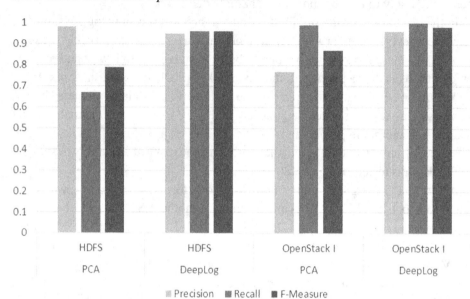

Due to their design, LSTM models naturally capture time-dependent relationships by processing sequences of events directly. They don't require feature extraction like the other ML and DL techniques do, lessening the data preprocessing workload. Experiments by Du et al. (2017) have also shown LSTM models to have relatively stable performance that is not very sensitive to changes in the hyperprameter or the characteristics of the dataset. Together, these factors make LSTM models great as general-purpose log analysis tools (Du et al., 2017).

Another benefit of LSTM models is that they process individual events (rather than feature vectors extracted from multiple events) in the detection phase. So when an anomaly does arise, it is easier for an operator to pin-point the events that caused the issue (Du et al., 2017).

The weakness of LSTM models, and RNNs in general, is that they require longer training time when compared to other deep learning models (Malaiya et al., 2018). The sequential processing of events makes LSTM models less parallelizable. Fortunately, once the model is trained, the time required for prediction in the detection phase is relatively short. Du et al. (2017) observed a prediction time of about 1 milliseconds per log entry in their experiments.

CONCLUSION

Logging is a useful practice that has been adopted in various fields to record events of interest for future reference. Logging in computer systems is unique in that log messages are automatically generated by computer programs, leading to significantly larger volume of logs. This chapter covered the basics of log collection and management, then focused on different log analysis techniques.

Rule-based systems use a rule-base (library of rules) to evaluate log events. These systems are effective in detecting known issues for which there are well-defined rules in the rule-base. They have been and will likely continue to be used in computing systems due to their simplicity and effectiveness. Where these systems fall short is in detecting new or unknown issues. This is where anomaly detection techniques come in.

Anomaly detection looks for unusual or abnormal behaviors that are different from the expected and normal behaviors. This goal can be achieved by many statistical, machine learning, or deep learning techniques. In this chapter, the authors surveyed many of these techniques. The authors started with simple techniques modeling system features as normal distributions, then moved on to machine learning techniques that combine statistics with an algorithmic approach such as k-Means and PCA, and finally moved on to novel deep learning techniques including MLPs and RNNs. The authors described the concepts behind each technique, their application to log analysis, as well as their strengths and weaknesses.

The experiment results by Malaiya et al. (2018) and Du et al. (2017) showed that deep learning models are more accurate than machine learning models in general. Furthermore, of the different types of deep learning models, RNNs such as LSTM produced the best results.

FUTURE RESEARCH DIRECTIONS

As we've seen in previous sections, deep learning models outperform traditional machine learning techniques for log analysis. Of deep learning techniques, RNN performed best due to its ability to model sequences, such as time-series data (Malaiya et al., 2018). Therefore, it is expected that the emerging trend will be more application of RNN models to log analysis, or the combination of RNN with other techniques.

Another challenge and research area in log anomaly-detection is the handling of *concept drift*, which is the change or drift of normal system behavior over time. An effective log analysis system needs to be able to adapt to this change or drift in system behavior in an online fashion (Debnath et al., 2018). This can be achieved by periodically retraining the model on new data. However, retraining can be inefficient and result in high overhead if the model must be trained from scratch each time. Ideally, the system would be able to make use of learning from past training and incrementally update itself based on new data efficiently. The feasibility of such incremental update mechanisms depends on the machine learning or deep learning technique used. Researchers have proposed incremental update mechanisms for machine learning techniques (Yi, Wu, & Xu, 2011) as well as deep learning techniques (Tsaih, Huang, Lian, & Huang, 2018).

The other emerging trend is the application of log analysis to logs from additional types of devices. A primary example of this includes Internet of Things (IoT), which introduces a large number of web-enabled devices such as smart home appliances. The research company Gartner projects that there will be over 20 billion IoT devices worldwide by 2020 (Meulen, 2017). Logging capabilities are often incorporated into these devices, resulting in large increases in the volume and diversity of logs. A great deal of these logs are already being analyzed for security and business purposes. For example, Ahmadon, Yamaguchi, Saon, & Mahamad (2017) proposed an approach to detect malicious activities through the analysis of the event logs of IoT services. There will be greater incentive for related research as more devices are added.

REFERENCES

Ahmadon, M., Yamaguchi, S., Saon, S., & Mahamad, A. (2017). On Service Security Analysis for Event Log of IoT System Based on Data Petri Net. *2017 IEEE International Symposium on Consumer Electronics (ISCE)*. 10.1109/ISCE.2017.8355531

Chuvakin, A., Schmidt, K., & Philips, C. (2013). *Logging and Log Management: The Authoritative Guide to Understanding the Concepts Surrounding Logging and Log Management*. Waltham, MA: Elsevier.

Debnath, B., Solaimani, M., Gulzar, M., Arora, N., Lumezanu, C., Xu, J., . . . Khan, L. (2018). LogLens: A Real-Time Log Analysis System. *IEEE 38th International Conference on Distributed Computing Systems (ICDCS)*.

Du, M., Li, F., Zheng, G., & Srikumar, V. (2017). DeepLog: Anomaly Detection and Diagnosis from System Logs through Deep Learning. In *Proceedings of the 2017 ACM SIGSAC Conference on Computer and Communications Security* (pp. 1285-1298). Dallas, TX: ACM. 10.1145/3133956.3134015

Goodfellow, I., Bengio, Y., & Courville, A. (2016). *Deep Learning*. MIT Press.

Han, J., Kamber, M., & Pei, J. (2012). *Data Mining: Concepts and Techniques* (3rd ed.). Waltham, MA: Elsevier.

He, S., Zhu, J., He, P., & Lyu, M. (2016). Experience Report: System Log Analysis for Anomaly Detection. *IEEE 27th International Symposium on Software Reliability Engineering (ISSRE)*.

Hochreiter, S., & Schmidhuber, J. (1997). Long Short-Term Memory. *Neural Computation, 9*(8), 1735–1780. doi:10.1162/neco.1997.9.8.1735 PMID:9377276

Kussul, N., & Skakun, S. (2005). Intelligent System for Users' Activity Monitoring in Computer Networks. Intelligent Data Acquisition and Advanced Computing Systems: Technology and Applications Conference.

Layman, L., Diffo, S., & Zazworka, N. (2014). Human Factors in Webserver Log File Analysis: A Controlled Experiment on Investigating Malicious Activity. *Proceedings of the 2014 Symposium and Bootcamp on the Science of Security.* 10.1145/2600176.2600185

Lima, M., Zarpelao, B., Sampaio, L., Rodrigues, J., Abrao, T., & Proenca, M. (2010). Anomaly Detection Using Baseline and K-Means Clustering. *International Conference on Software, Telecommunications and Computer Networks (SoftCOM)*.

Liu, B. (2011). *Web Data Mining: Exploring Hyperlinks, Contents, and Usage Data* (2nd ed.). Chicago, IL: Springer. doi:10.1007/978-3-642-19460-3

Malaiya, R., Kwon, D., Kim, J., Suh, S., Kim, H., & Kim, I. (2018). An Empirical Evaluation of Deep Learning for Network Anomaly Detection. *International Conference on Computing, Networking and Communications (ICNC).* 10.1109/ICCNC.2018.8390278

Meulen, R. (2017). *Gartner Says 8.4 Billion Connected "Things" Will Be in Use in 2017, Up 31 Percent From 2016.* Retrieved from https://www.gartner.com/en/newsroom/press-releases/2017-02-07-gartner-says-8-billion-connected-things-will-be-in-use-in-2017-up-31-percent-from-2016

Olah, C. (2015). *Understanding LSTM Networks.* Retrieved from http://colah.github.io/posts/2015-08-Understanding-LSTMs/

Powell, V. (n.d.). *Principal Component Analysis: Explained Visually.* Retrieved from http://setosa.io/ev/principal-component-analysis/

Russell, S., & Norvig, P. (2010). *Artificial Intelligence: A Modern Approach* (3rd ed.). Upper Saddle River, NJ: Pearson Education.

Stamp, M. (2006). *Information Security: Principles and Practice.* Hoboken, NJ: John Wiley & Sons.

Tsaih, R., Huang, S., Lian, M., & Huang, Y. (2018). ANN Mechanism for Network Traffic Anomaly Detection in the Concept Drifting Environment. *IEEE Conference on Dependable and Secure Computing (DSC).*

Xu, W., Huang, L., Fox, A., Patterson, D., & Jordan, M. (2009). Detecting Large-Scale System Problems by Mining Console Logs. *Proceedings of ACM Symposium on Operating Systems Principles (SOSP)* (pp. 117-132). Big Sky, MT: ACM. 10.1145/1629575.1629587

Yi, Y., Wu, J., & Xu, W. (2011). Incremental SVM Based on Reserved Set for Network Intrusion Detection. Expert Systems with Applications: An International Journal.

Yin, C., Zhang, S., Wang, J., & Kim, J. (2015). An Improved K-Means Using in Anomaly Detection. *First International Conference on Computational Intelligence Theory, Systems and Applications (CCITSA).* 10.1109/CCITSA.2015.11

ADDITIONAL READING

Chuvakin, A., Schmidt, K., & Philips, C. (2013). *Logging and Log Management: The Authoritative Guide to Understanding the Concepts Surrounding Logging and Log Management.* Waltham, MA: Elsevier.

Goodfellow, I., Bengio, Y., & Courville, A. (2016). *Deep Learning.* MIT Press.

Han, J., Kamber, M., & Pei, J. (2012). *Data Mining: Concepts and Techniques* (3rd ed.). Waltham, MA: Elsevier.

Liu, B. (2011). *Web Data Mining: Exploring Hyperlinks, Contents, and Usage Data* (2nd ed.). Chicago, IL: Springer. doi:10.1007/978-3-642-19460-3

Olah, C. (2015). *Understanding LSTM Networks.* Retrieved from http://colah.github.io/posts/2015-08-Understanding-LSTMs/

Stamp, M. (2006). *Information Security: Principles and Practice.* Hoboken, NJ: John Wiley & Sons.

KEY TERMS AND DEFINITIONS

Anomaly Detection: Analyzing data to detect unusual or abnormal behavior.

Artificial Neural Networks (ANN): Computing systems that use networks of interconnected nodes to process and gain knowledge from training data, then apply the knowledge to make predictions.

Event Correlation: Looking across different events to extract global insights based on their relationships.

K-Means: Machine learning technique that identifies groups/clusters of data points that are similar to each other.

Log Analysis: The analysis of Logs to extract useful information for troubleshooting, monitoring, auditing, and other purposes.

Long Short-Term Memory (LSTM): Type of RNN that incorporates multiplicative gates that allows the network to have long- and short-term memory.

Multilayer Perception (MLP): Class of ANN that are feedforward and fully connected in construction.

Principle Component Analysis (PCA): Machine learning technique that transform the data to identify important relationships and reduce the dimensionality of the data.

Recurrent Neural Networks (RNN): Class of ANN that have recurrent connections that allow the network to maintain internal state/memory.

Rules Engine: Software that allow the user to specify rules in a library (known as a rule-set), which the software then applies for various purposes. In the context of log analysis, rules engines use the rules to evaluate log events and take appropriate actions.

Chapter 8
Password–Less Authentication:
Methods for User Verification and Identification to Login Securely Over Remote Sites

Rahul Singh Chowhan
iD https://orcid.org/0000-0001-6567-4979
Agriculture University Jodhpur, India

Rohit Tanwar
University of Petroleum and Energy Studies, India

ABSTRACT

Over the years, passwords have been our safeguards by acting to prevent one's data from unauthorized access. With the advancement of technologies, the way we have been using passwords has changed and transformed into much secure yet more user friendly than they were ever been in the past. However, the vulnerabilities identified and observed in this traditional system has motivated industry and researchers to find some alternate where there is no threat like stealing, hacking, and cracking of password. This chapter discusses the major developed password-less authentication techniques in detail and also puts an effort to explain the in-depth details along with the working principle of each of the technique through a use-case diagram. It would be of great benefit and contribution to the callow trying to explore research opportunities in this area.

DOI: 10.4018/978-1-5225-8100-0.ch008

INTRODUCTION

Over the years, passwords have been stolen, cracked and hacked. Fraudulent agencies can buy user information and credentials online on social media sites. Many cases have been seen worldwide like Facebook data leak, Yahoo Security Breach, LinkedIn Data Breach, DropBox User Accounts leak etc. Another reason could be to increase in the variety of applications and platforms that could force the user to remember more and more passwords (Cortopassi, M., Edward, E., 2013). As technology and its users keep on increasing with the demand-branding, publicity and efficiency of the application, there is an increase in secure channels to communicate and store passwords. Although, password-based login is more prevalent in today's time but because of the drastic increase in internet-connected devices and user's possessing more online accounts than ever before has made password-less authentication a more relevant alternative for secure logging-in to online accounts (Rabkin, A., 2008). It becomes difficult to memorize passwords and that would lead users to keep one password for most of the application causing them prone to hackers. This is the reason that could actively lead to an increase in security breaches and easier for hackers to capture data. This has also fostered the applications that keep a store of all user accounts and passwords associated with respective accounts that user uses locally. At this end, the password management scheme seems to be a promising and reliable factor to store tricky passwords for accessing cross-platform systems with single sign-on. The layman user thinks them as time savvy and less tedious as they keep the bulky password at one place. But the user does not understand how these applications might work behind the walls in the backend to share their sensitive information across the internet. Though, along with acceptance of terms during the installation process or registration process, the user may unknowingly allow the application to share the sensitive information (Luke, Hok-Sum H., Matthew W. T., 2015). This time the trustworthiness of a user can be tested by blindly allowing these kinds of applications to access secure accounts. The password occurrence per user stays common and relative to each other which would also cause hackers to guess passwords by using hit and try the method. This hit-and-trial method causes severe problems like gaining remote access to obtain user information stored either on a client machine or server. After facing and accepting all the challenges we are at the step of no more password breaches with a promise of more secure authentication and no password memorization at all (Chiasson, S., Elizabeth, S., Alain, F., Robert, B., Paul, C. V. O., 2012). Passwordless authentication is a critical investment on security which serves various benefits as under:

1. **Enhanced User Experience:** New age users need notto remember the use of puzzles and questions like "What is your first pet dog name?", "Which was your high school?" etc. This not only reduces the signup time but also eliminates users to go through a tedious registration process for new apps. They have much interactive interface and facilities than the password-based authentication.

2. **Improved Security:** With no use of passwords we have more secure improvised security with zero passwords to remember.

Security is the main dimension of any software development application which involves various processes of authentication, verification, authorization, and integrity. The user, as well as developer of the application, could not overlook these aspects while using and developing the application. Increase in security breaches in the past decade has forced the developers to come up with more secure strategies and authentication handling mechanisms (Mahaffey, K. P., et.al. 2016). One such authentication mechanism is the use of passwordless authentication. In this authentication service, users are no more required to remember tricky passwords for different applications that they use in daily life. This type of authentication does not require passwords to login into any application. The password-less authentication has become the talk of town making passwords obsolete day by day. This new age enterprise model ensures the effortless security based on these enhanced user authentication mechanism (Fox, A., Steven, D., G.,1996). This kind of authentication showcases various options for securely logging for users.

DIFFERENT TECHNIQUES OF PASSWORD-LESS AUTHENTICATION

The working of various password-less authentication modalities are as follows:

1. **Magic Link Login Authentication:** In day-to-day life user logins to various accounts so rather than asking users to enter passwords, just get the username to generate a transient authorization code for the session-full login. Simultaneously, this code is acknowledged to mail id for the associated username provided by the user. Later with the time-bound access of magic link user needs to click the link to verify its identity for accessing the account based services. In the back end, once the user logs in to the associated account this authorization code is then exchanged with the session key to let user access account as long the session times out (Lee, W., Simon, SH., P., Chasing, L., Jinho, K.,

Byeong-Soo J., 2014). This same key is stored at user device as well. The magic link authentication down the line believes that mail server is safe enough to authenticate the user's identity as the user cannot regulate the backend security at mail databases. The general flow in passwordless magic link login system may follow as:

a.   The user is asked to fill up the username
b.   On click of submitting, two API calls are used
c.   First to begin the verification process that generates an authentication code, this will be sent to the registered user's mail id.
d.   Second to compare the authentication code previously generated and another which is obtained with user's navigation from mail to app/account.
e.   After a successful comparison, the user is navigated to the app and the matched authentication code is exchanged with the secure session key for longer access to the account.
f.   The magic link is now disabled.

Figure 1. API Call 1 (for authentication key)

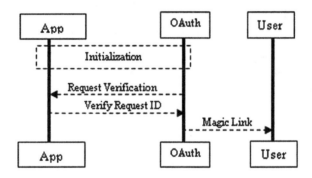

Figure 2. API Call 2 (to get authenticated)

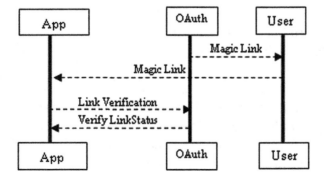

Figure 3. Working: Getting magic authentication link on Email

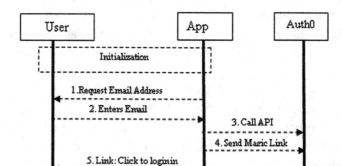

Figure 4. Working: Authenticated login to the user account

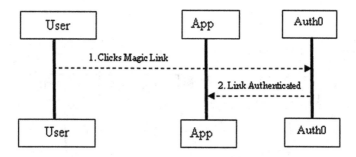

Now with this scheme of authentication has a challenge to face if mailbox or mail client of the user is compromised. This will open the access to the user account to someone using the mail id (Popp, N., David, M., Loren, H., 2011). Then with the help of a magic link, someone else might access the account as long as the validity of the link exists. However, there persist many solutions to this problem, which are as shown in Box 1.

2.   **Token-Based Login (OTP Based):** The OTP is the most commonly used form of two-factor authentication for secure access of network. At cryptographic level, the multi-factor authentication differentiates this terminology into "what user has", "what user knows" and "what user is" in terms of cryptography (Bychkov, E., 2012).The token-based mechanism works similar to magic link authentication. In this rather than using email id link the registered mobile number of the user is used to authenticate it. As well as they provide an authentication token which is solely enough to authenticate the user. Nowadays,

Box 1.

Solution 1: Maintain a blacklist of the magic link that can be canceled by the app owner.
Solution 2: Forming a Revision Index, i.e. the link with the latest revision would be allowed.
Solution 3: Short span of Magic Links i.e. quick expiry say 5 minutes to login.
Solution 4: Rebuild the link in two pieces so user logging would be authenticated by the first split sent to mail and second one sent in response to the request by the user.

many organizations make use of OTP in the combination of Username & Static password. As the name is One Time Password that means it is valid for single time interaction, single session, a single transaction or single authentication. These are more secure than the user created static password mainly for two reasons (M'Raihi, D., Salah M., Mingliang P., Johan R., 2011). First, they are different and dynamic for different users that mean the brute force attacker has more guesses to break and second they are of short life span validity minimizing the risk of replay attacks. Cryptographically, they are tokens generated by a counter with a combination of token key and a current counter value which is unique and secure notion followed by server and token manager in synchronization. They can not only be used for simple user account login but also for 3D secure transactions and payment gateways, which often requires a tricky combination of alphanumeric and special characters with various validation. Like a typical 3D secure transaction would ask its user to create a password with at least two special characters, a minimum of four digits and six characters which may not include spaces and tabs (Steeves, D., J., Mwende, W., S., 2007). There is absolutely no limit to adding up a validation to not allow compromise on user information. But it becomes difficult for a layman user to remember this kind of password which may end up in a recreation of it in next login. The static passwords are more gullible to Password Guessing Probability (PGP) attack than One Time Password or One Time Code. This has become the most common password-less authentication service to protect various network devices like Firewall, Router, Switches VPN Users, etc. that enterprises preferto use. The One Time Password is also known as Pass Code as it helps the user to get authenticated by Authentication Manager (Curry, S., M., Donald, W., Loomis, Christopher W., F., 2000). In general, an Authentication Manager is the software that generates and validates a user before allowing login securely. It matches the OTP sent to the user and the Pass Code saved with it. The steps for the general flow of token-based authentication are as follows:

a.   User visits the login page& inputs the registered phone number and submits

b. A token, 4 or 6 digit PIN, is generated by the authenticator at server side which is associated with the registered number after submission

c. On submit, Open Authorization (OAuth) will send the token to the user's mobile number.

d. The GUI on app/site asks for that 4 or 6 digit PIN to be filled by the user.

e. Second API call compared the previously generated PIN with the PIN entered by the user on GUI.

f. If both PIN matches, the user is securely authenticated to access the associated account.

OTP Tokens can be generated synchronously as well as asynchronously based on which they are classified into two main categories.

3. **Certificate/Counter Based Security Token Authentication:** This type of authentication is also called as HOTP (Hash Message Authentication Code based One Time Password). It happens between the client's token and token generated by the authentication server. Regardless of Time Based Token i.e.

Figure 5. API Calling in Token Based Login

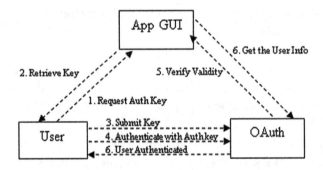

Figure 6. Working: Requesting and Obtaining Token on Registered Phone Number

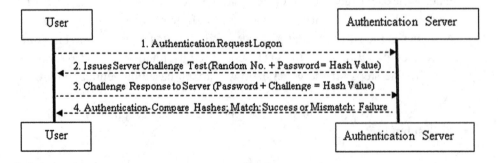

Figure 7. New User: Authenticating new user by adding it to the connection

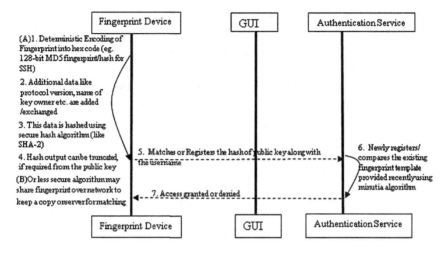

Figure 8. Registered User: Authenticating previously registered user

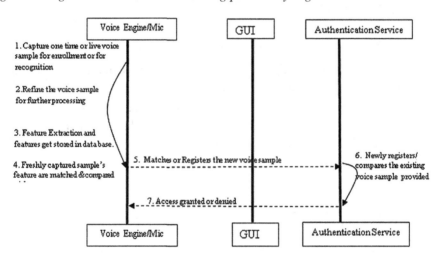

TOTP which changes on its own after every few seconds, this requires the user to tap on token button asking for next token in counter (Shaw, J., Richard H., Mike A., James N., et. al., 2003). In this, the server maintains the windows of token numbers so if the user taps twice or enter an expired token it will not be authenticated (Giobbi, John J., David, L., B., and Fred S. H., 2011). The major drawback of counter-based token authentication is that if the user taps multiple times on the token generator button and is not authenticated for any of them. Then it may accidentally manage to get over the window of numbers and user has to start all over again by resetting the whole authentication process. The changing value of the counter called a moving factor, which gets incremented after usage. Both the parties that are getting involved in the authentication process needs to remember the last used value of the counter. In this process, the client and server may get desynchronized due to a network issue. If the client has sent the token and incremented its moving value and due to an error in connection if the server never received it, then on a comparison of counter value at the server will result into the process of manual resynchronization (Khan, M. A., Hasan, A., 2008). As there is a lack of single synchronization technique, the communication takes place in unidirectional only.

Mathematical notation of HOTP Cryptographic function:

$$HOTP(K,C) = Trunc\left(HMAC - SHA - 2(K,C)\right)$$

where; K is shared key and C is counter

4. **Time-Based OTP (TOTP):** This is a transient passcode that authenticates the access of user accounts. It is temporary in nature and is generated using the current time as one of its parameters making each password unique and strong to compromise. In general, they are used with traditional login system of username/password for two-factor authentication. Even though the username

Figure 9. Server Verifying the User using HOTP

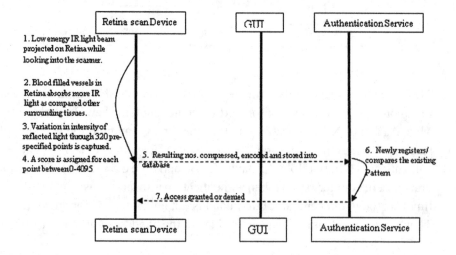

and password of user accounts are compromised the TOTP will unauthenticated the false access to accounts (M'Raihi, D., Mihir B., Frank H., David N., Ohad, R., 2005). There exist many ways of TOTP which include: Hardware Security Token, Mobile apps, and Text Messages. The TOTP is often 8 digits long numeric code valid for 30 or 60 seconds and changes frequently that means the brute force attacker will almost run out of time to break through new credentials every time (Conklin, A., Glenn, D., Diane, W., 2004). It is generated from a shared Private Key and the Current Time Zone using HMAC (Hash-based Message Authentication Code) so they are also called as HOTP (M'Raihi, D., Salah, M., Mingliang, P., Johan, R., 2011). There is a slight difference between HOTP and TOTP which are stated in Table 1.

Mathematical notation of TOTP Cryptographic function:

$$TOTP = HOTP\left(K, T\right)$$

$$HOTP\left(K, T\right) = Trunc\left(HMAC - SHA - 2\left(K, T\right)\right);$$

where, K is shared key and T is no. of time stamps between initial courier time $\left(T_0\right)$ and current system time

T is given as:

$$T = floor\left(\left(current\ time - T_0\right) / X\right);$$

where X is time stamps (sec)

$$By\ default\ T_0 = 0\ and\ X = 30sec$$

The only condition with TOTP is that the device generating TOTP must be in sync with Standard Time Zone of that area. It is generated on both ends i.e. on the server and client app so it is always available with the client allowing them to enter Zero Wait Zone eliminating the network dependency in case of OTP (Soare, C., Andrei., 2012).So in this to utilize the higher level of security the user has to be fast enough to enter the generated token before the timer times out. It is also called a combination of "something you know" i.e. Password and "Something you have" i.e. a device with TOTP. The diagram below shows the communication mechanism and authentication of the user using TOTP:

Table 1. Comparison of HOTP and TOTP

HOTP (HMAC Based One Time Password)	TOTP (Time-Based One Time Password)
Relies on Shared Key and Counter (moving factor), Counter values are generated based on the shared key	Relies on Share key and Counter but counter values changes with the function of time (every 30 or 60 sec)
Event-basedOTP algorithm is used i.e. the moving factor is an event counter	Time-based OTP algorithm is used i.e. short-lived OTP values for enhanced security are used
The counter is subsequently incremented whenever a new OTP is generated	The counter values are incremented with a timeout of an OTP
Pass Token is valid for an unknown amount of time	Passcodes are valid only for short duration of time (can also be called as Extension of HOTP that involved time)
Requires more maintenance but no synchronization between user device and server	Requires less maintenance but high-level synchronization because of the factor of time
Published as IETF RFC 4226	Published as RFC 6238
An example can be Yubikey, Google Authenticator	Example Google Authenticator
Less Expensive as the counter is cheaper to implement for authentication purpose	More Expensive because accurate time devices are required for authentication

Figure 10. Verifying the user using TOTP authentication

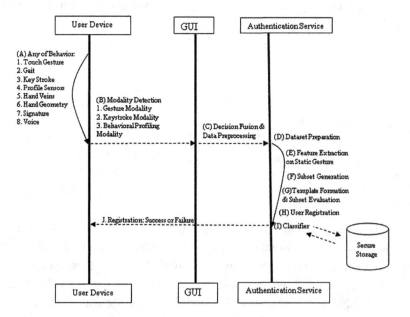

There exist few drawbacks with both the techniques like a major issue with HOTP is user can get locked in the self loop by multiple tapping of a token button. Another problem could be related to session validity, the authentication is valid until the user uses the token. Once it is used it will not be reused again. Whereas in the case of TOTP, the session is only authentic until the timer times out after that user needs to re-sync the whole process again. Both of the protocols suffer from the common issue of storing the shared secret key on the user's local device that makes it susceptible to attacks (Abukeshipa, A., SM., Tawfiq, SM., B.,2014).

5. **Challenge Response Authentication Mechanism (CRAM):** This is the family of protocols with characteristics of one entity sending challenges to other entity. On receiving the appropriate answer from the second entity the authenticated access is allowed. It is a way to prove the user identity over an insecure communication channel without giving any information to an eavesdropper.It is the two-step scheme for verifying the user connected in the network from HTTP that makes use of one way hash functions which makes it infeasible to find the input of function by using a generated hash. The two steps include user authentication and digest authentication. Like most smart card systems make use of CRAM which requires two entities: user smart card and password (Mizrah, L., L., 2011). Another example is the use of CAPTCHA which recognizes the auto-registration, prevent spam and determine human inputs. In cryptography, Key Agreement mechanism like CRAM-MD5, Secure Shell CRS, Zero-Knowledge Proofs based on RSA are secure CRAM. The major drawback with this scheme is sending of same challenge more than once. If the server throws the same challenges again and again then it will make the system gullible to replay attack (Rhee, K., Jin Kwak, S. K., and Dongho W., 2005). The attacker might record the hash of the previous authentication and can simply replay the hash to gain the secure access without knowing the password. The working of CRAM is shown in a diagram below:

6. **Fingerprint/ Thumbprint Authentication:** Most of the app developers expect users to form a complicated combination of passwords that are unique and un-guessable on first sight. Though it is need of time to having an easy yet secure authentication pass-code mechanism. This must be secure, unique, cannot be forged, and provides quicker access to app or service. One such way is using a fingerprint of the user that is easy to do as everyone possesses a smart-phone and also nothing to remember like in password-based authentication schemes (Morales, A., Travieso, C. M., Ferrer, M. A., Alonso, J., B., 2011).

Figure 11. Authentication using CRAM

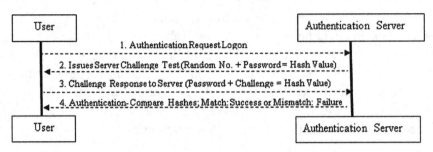

Authentication using fingerprints is not a novice concept it has been in the market a couple of decades before in laptop, automatic suitcase and electronic doors etc. but now it has been fanned more with the prevalence of mobile gadgets. Though fingerprint scanning is a better option for identity verification and detecting duplicates but it is not secure enough for proving the authenticity of a person. In the year 2013, Apple released TouchID based fingerprint authentication for Apple iPhone users that have boomed the biometric market with similar features in Android and Samsung. For Apple users, it could be used for two purposes of unlocking the phone and authenticate app store. This form of authentication allows to user scan fingerprint by placing their finger on a mobile device. The verification process happens by analysis of special features of prints called minutiae. This is basically the lines on fingers that terminates and bifurcate by creating distances and angles (Yi, C., Dass, S. C., Jain A. K., 2005). This measurement is unique for every person and based on which a unique key pair is generated locally using a pattern matching algorithm on the user device while at the server a new user is registered that has a mapping with a unique key pair of fingerprint. A new session is initialized and the user gets authenticated to securely login to an associated account.

The fingerprint of the user is never sent over the network. In fact, when a user signs up using fingerprint authentication, a unique key pair, secret key, and the public key is generated on the user device. The user is created, the public key is assigned and the private key is saved in a key store and all this happens locally on the user device. Not only this much but fingerprints are also used biometric attendance system, a collection of forensic evidence during criminal investigations, to control access to devices like smartphone, laptops etc. The most important part of fingerprint scanning and authentication is quality control which decides the ridges, valleys and required details of minutia points on fingers accurately. There is no problem called as forgetting passwords, creating a strong combination of alphanumeric cum special characters and as well as no fear of losing them. Swipe-and-PIN was also a promising alternative for payments using e-wallets (Rassan, Iehab, AL., Shaher,

H. Al., 2013). These used magnetic strips to store user's personal data like credit card number and the user is asked to swipe the card and insert the associated PIN to complete the transaction. But many of security breaches with the merchants have been recorded on checkout machines in the US. This has coined more secure and new technologies like Chip-and-PIN, Chip-and-Signature, and Swipe-and-PIN.

Any key management technique involves key generation, key modification and key sharing for establishing in secure message communication over a network.

The process is shown in Fig(12).

7. **Authentication Using Face Recognition (Selfie Authentication Service):** User is capable ofopen the camera application within the device which captures the current face in real-time. This captured face is then compared with the file stored in the device media to get the genuine access of the device. Upcoming face recognition technologies are capable of providing high-level of confidence to oversee facial variances due to age effects like wrinkles and more (Avital, A. 2017). This enables more live detection and determination of the user in case if the user is using a steady photo for insecure login. The facial recognition software tools recognize the facial data points and liveness detection for accurate authentication. Face recognition algorithms have evolved from primitive models to advanced models making use of Neural Networks, Machine Learning algorithm and many more for authentic unlocking of the device or access of account. The access control system of face authentication involves capturing devices, image processing entity and recognition algorithm (Bowyer, Kevin W., Chang, K., Flynn, P., 2006).

Nowadays, the high definition spatial cameras are used for surveillance at public places. They can help in tracking fugitive criminals, missing people and observing many suspicious activities without even been noticed by the subject. Thought they work differently in comparison with the authenticated facial login mechanism (Atick, Joseph, J., Griffin P., A., Redlich, A. N., 2000). Many authentication devices can configure with multiple authentication feature, this way make use of additionally added security layer for a much advanced level of user authentication.

Working Principle of Face Recognition Authentication

    a.   Draw connections of data points of the face on highly accurate, secure and sophisticated machine learning algorithm to generate the biometric template of the face.

    b.   Compares the newly generated biometric template with the previously stored template.

    c.   Securely authenticates the match based on accuracy score.

Figure 12. Authentication using Thumb/Fingerprint

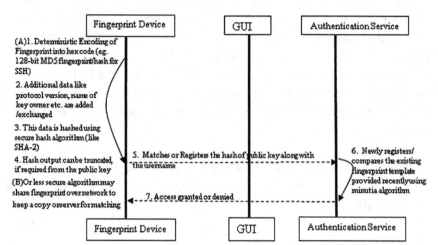

8. **Voice Recognition Based Login (Voiceprints):** It does the comparison of voiceprints stored on the device with the voice of a person trying to authenticate it. Due to rapid growth in the fraud cases related to the banking system wherein the victim is contacted through telephonic means, banking systems are investing more in authentication using voice. It is because of the higher accuracy being achieved in this technology day by day that most of the security-related problems are looking for a solution using voice recognition (Hunt, A. K., Thomas B. S., 1992).

Speech recognition and Speaker recognition are two subsets of voice recognition having slightly different focused domains. While speaker recognition is intended to authenticate the human using its voice features but not interested in understanding the meaning of words uttered by the person but the speech recognition does this also wherein the words are further used to command or to control something (Yuanchun, shi., Xie, Weikai., Xu, Guangyou., Shi, Runting., et. al. 2003).

Voice Recognition is further divided into two types; recognition and identification as follows:

a. **Recognition:** It is thetermusedwhen an unknown user raises a claim in his/her voice using some ID. The system then matches the voice prints of newly received samples with the voice prints saved in a database for that particular ID and assigns some matching score. If the score is above some threshold value then the claim is accepted and authentication is granted otherwise not. Since the comparison is done with a single database entry, this process is comparatively easy.

b. **Identification:** It is the term used when an unknown person raises a claim in his/her voice and with no ID. The challenge is to identify the person now out of all the registered users if a successful match is there. The given voice sample's prints are matched with existing voice prints and assigned some matching scores. All the scores are then ranked. The topmost ranker's matching score is compared for threshold restrictions. If succeeded then it is authenticated otherwise not (Dimitri K., Maes S. H. 2000).

here are so many factors that affect the voice sample and hence increase challenges of the voice-based authentication system; like surrounding noise, physical and mental fitness of the speaker, accent of the speaker and so on. Because of these challenges, it is the area of interest of many of the researches these days.

9. **Retinal Scan Authentication:** The human retina that is situated in the back segment of the eye is a thin tissue made out of neuralcells. In light of the unpredictable structure of the vessels that supply the retina with blood, every individual's retina is one of a kind. The system of veins in the retina is complex to the point that even indistinguishable twins don't share a comparative example (Shanley, C. W., Jachimowicz, K., Lebby, M., S., 1994). It might happen that retinal patterns got modified in few diseases like as of diabetes, glaucoma or retinal degenerative issue; the retina normally stays unaltered from birth until death. Retina-scan innovation first launched in the 1980s is outstanding however most likely it has gain very low popularity of all the biometric advancements.

Figure 13. Authentication using Voice

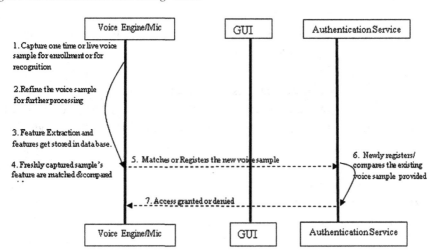

Moreover, the retina-examine technique is still in a model improvement stage and still financially inaccessible.

Retinal Scan technology depends on the vein design in the retina of the eye. The rule behind the innovation is that the veins at the retina give a one of a kind example, which might be utilized as a non-alterable identifier. Since infrared waves are assimilated quicker by veins in the retina than by encompassing tissue, it is utilized to enlighten the eye retina. Investigation of the upgraded retinal vein picture at that point happens to discover trademark designs. Retina-check gadgets are utilized solely for physical access applications and are normally utilized in situations that require high degrees of security, for example, abnormal state government military needs.

Unlike this in iris authentication technology, the identification task is completed by get-together at least one itemized pictures of the eye with a modern, high-goals advanced camera at unmistakable or infrared (IR) wavelengths, and after that utilizing a specific PC program called a matching engine to compare the subject's iris image and the pictures already stored in a database. The engine is able to compare a large number of pictures every second with a level of accuracy similar to customary fingerprinting or computerized finger scanning (Britz, D., Robert R. M., 2012).

All together for iris recognition to give precise and reliable outcomes, the subject must be inside a couple of meters of the camera. Some control systems must be executed to guarantee that the caught picture is a genuine face, not a fantastic photo. The encompassing lighting must not create reflections from the cornea (the gleaming external surface of the eyeball) that darken any piece of the iris. The subject must stay stationary, or almost stationary, regarding the camera, and must not be antagonistic to the procedure. Certain kinds of contact focal points and glasses can darken the iris design (Nichols, T J., Thompson, D. L., 2005).

Iris scan likewise discovered across the board selection in Government's drives relating to Aadhar and UID. Aadhar is being made required by numerous associations. As Aadhaar enlistments cross the 1 billion checks, the utilization of Aadhaar for occupant verification and eKYC is relied upon to quickly spread widely in services provided by the government and private firms. The requirement for a biometric methodology that works dependably is basic for the utilization of Aadhaar, a key segment of the administration's Janardhan Aadhaar Mobile (JAM) drove activities. The iris is generally acknowledged as a more secure and dependable methodology to recognize and confirm individuals. The iris works over a more extensive populace, overall age gatherings and occupations. This verification benefit empowers an Aadhaar holder to distinguish himself/herself utilizing the picture of his iris picture. It is expected by the market experts that retina/iris scan technology market is set to grow from \$676.6 million in 2016 to \$4.1 billion by 2025. In future everything starting from your bank account to air travel ticket will get unlocked through this

Figure 14. Authentication using Retina

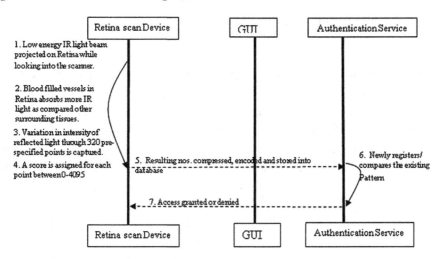

technique. However few smart-phones are coming enabled with technology at present (Bolle, R. M., Sharon L. N., et. al. 2004).

10. **Behavioral Biometric Authentication:** Behavioral biometrics is described as identifying, measuring and recording some patterns in the behavior of human and utilizing them to authenticate that human in real time or because of retrospection. Instead of concentrating on a result of an activity, it centers on how a client does the predetermined action. For example, while asking for entering username and password, it doesn't look whether they are entered correctly or not rather it notices their typing speed (slow or fast), whether the mouse was used for changing cases or with the keyboard (Banerjee, S. P., Damon L. W., 2012). Digital devices like cell phones, tablets and wearable gadgets, etc. are the prime data providing sources of behavioral biometrics. Improvement in size, connectivity, efficiency and configuration of sensors is supportive in the growth of this technology. Most of the consumer devices are equipped with advanced sensors. It is very easy to get data related to the behavior of an individual through a smartphone or similar devices having accelerometer and gyrometer (Wayman, J., Jain, A., Maltoni, D., Dario M., 2005). Advanced and smart algorithms are available now that can analyze, interpret and utilize that data to fight against fraud authentication. For additional security and preventing biometric data from mischievous attacks, it is preferred to encrypt the acquired data before storing into database. After data acquisition, few points are nominated as match points by specific software. These points are then processed by an algorithm which is responsible for converting the

extent of matching in some numerical value. The value stored in the database is compared with the input provided by the user for biometric and then the decision of granting or denying access is taken accordingly.

The most positive fact in favor of this technology is its compatibility with the existing hardware. What it requires additionally is some specific software only. Because of this, the technique is cheaper and easy to deploy. The market is already in transition from traditional biometric system to behavioral biometrics. The leading market analyst Technavio expect that the annual growth rate of the behavioral biometric market will be 17.34% in the duration of 2016 to 2020. With the rise in deploying services on an online platform, the high demand for tamper-proof authentication technology is getting voice raised (Yuxin, M., Wong, D. S., Schlegel, R., 2012). It is the time now when moving from traditional authentication approaches like physical signatures to PIN and then to fingerprint recognition recently, behavioral biometrics seems to become as next popular part of our digital life in future.

CONCLUSION AND FUTURE SCOPE

This has increased the number of possibilities with new vendors to come up with a more secure and less tedious single point authentication solution for future customers.

Figure 15. Authentication using Behavior Biometrics

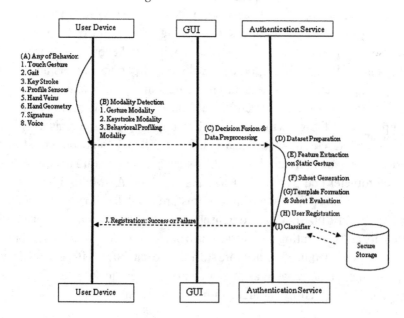

They can right away reach to sort out customer problems with their voice, face or the way they interact with their authentic devices. They may also be provided with multiple modalities like two-factor biometric authentications without any direct PINs and password sharing. These various credential free authentications without any interrogations and having to prove who they are, allow customers to be choice-free from the side of service providers. Due to its vast benefits, the market for face authentication, iris authentication, Voice authentication and etc. is expected to have a huge growth by 2021.

REFERENCES

Abukeshipa & Barhoom. (2014). *Implementing and Comprising of OTP Techniques (TOTP, HOTP, CROTP) to Prevent Replay Attack in RADIUS Protocol.* Academic Press.

Atick, J. J., Griffin, P. A., & Norman Redlich, A. (2000). *Continuous video monitoring using face recognition for access control.* U.S. Patent 6,111,517.

Avital, A. (2017). *Authentication using facial recognition.* U.S. Patent 9,547,763.

Banerjee, S. P., & Woodard, D. L. (2012). Biometric authentication and identification using keystroke dynamics: A survey. *Journal of Pattern Recognition Research, 7*(1), 116–139. doi:10.13176/11.427

Bolle, R. M., Nunes, S. L., Pankanti, S., Ratha, N. K., Smith, B. A., & Zimmerman, T. G. (2004). *Method for biometric-based authentication in wireless communication for access control.* U.S. Patent 6,819,219.

Bowyer, K. W., Chang, K., & Flynn, P. (2006). A survey of approaches and challenges in 3D and multi-modal 3D+ 2D face recognition. *Computer Vision and Image Understanding, 101*(1), 1–15. doi:10.1016/j.cviu.2005.05.005

Britz, D., & Miller, R. R. (2012). *Method and apparatus for eye-scan authentication using a liquid lens.* U.S. Patent 8,233,673.

Bychkov, E. (2012). *Extended one-time password method and apparatus.* U.S. Patent 8,132, 243.

Chen, Y., Dass, S. C., & Jain, A. K. (2005). Fingerprint quality indices for predicting authentication performance. In *International Conference on Audio-and Video-Based Biometric Person Authentication* (pp. 160-170). Springer. 10.1007/11527923_17

Chiasson, Stobert, Forget, Biddle, & Van Oorschot. (2012). Persuasive cued click-points: Design, implementation, and evaluation of a knowledge-based authentication mechanism. *IEEE Transactions on Dependable and Secure Computing, 9*(2), 222-235.

Conklin, D., & Walz. (2004). Password-based authentication: a system perspective. In *System Sciences*. Proceedings of the 37th Annual Hawaii International Conference on, 10.

Cortopassi, M., & Endejan, E. (2013). *Method and apparatus for using pressure information for improved computer controlled handwriting recognition data entry and user authentication.* U.S. Patent 8,488,885.

Curry, S. M., Loomis, D. W., & Fox, C. W. (2000). *Method, apparatus, system, and firmware for secure transactions.* U.S. Patent 6,105,013.

Fox, A., & Gribble, S. D. (1996). Security on the move: indirect authentication using Kerberos. *Proceedings of the 2nd annual international conference on Mobile computing and networking,* 155-164. 10.1145/236387.236439

Giobbi, J. J., Brown, D. L., & Hirt, F. S. (2011). *Personal digital key differentiation for secure transactions.* U.S. Patent 7,904,718.

Hunt, A. K., & Schalk, T. B. (1992). *Simultaneous speaker-independent voice recognition and verification over a telephone network.* U.S. Patent 5,127,043.

Kanevsky, D., & Maes, S. H. (2000). *Apparatus and methods for providing repetitive enrollment in a plurality of biometric recognition systems based on an initial enrollment.* U.S. Patent 6,092,192.

Khan, M. A., & Hasan, A. (2008). Pseudo-random number based authentication to counter denial of service attacks on 802.11. Wireless and Optical Communications Networks, 2008. In *WOCN'08. 5th IFIP International Conference.* (pp. 1-5). IEEE.

Lee, Park, Lim, Kim, & Jeong. (2014). Server authentication for blocking unapproved WOW access. *2014 International Conference on Big Data and Smart Computing (BIGCOMP),* 155-159. 10.1109/BIGCOMP.2014.6741427

Luke & Taylor. (2015). *Apparatus, method, and article for authentication, security and control of power storage devices, such as batteries.* U.S. Patent 9,182,244.

M'Raihi, Machani, Pei, & Rydell. (2011). *Top: Time-based one-time password algorithm.* No. RFC 6238.

M'Raihi, Machani, Pei, & Rydell. (2011). *TOTP: Time-based One-Time Password algorithm.* No. RFC 6238.

M'Raihi, D., Bellare, M., Hoornaert, F., Naccache, D., & Ranen, O. (2005). *An Hmac-based One-Time Password algorithm.* No. RFC 4226.

Mahaffey, Richardson, Salomon, Croy, Walker, Buck, … Golombek. (2016). *Multi-factor authentication and comprehensive login system for client-server networks.* U.S. Patent 9,374,369.

Meng, Y., Wong, D. S., & Schlegel, R. (2012). Touch gestures based biometric authentication scheme for touchscreen mobile phones. In *International Conference on Information Security and Cryptology* (pp. 331-350). Springer.

Mizrah, L. L. (2011). *Two-channel challenge-response authentication method in random partial shared secret recognition system.* U.S. Patent 8,006,300.

Morales, A., Travieso, C. M., Ferrer, M. A., & Alonso, J. B. (2011). Improved finger-knuckle-print authentication based on orientation enhancement. *Electronics Letters, 47*(6), 380–381. doi:10.1049/el.2011.0156

Nichols, T. J., & Thompson, D. L. (2005). *User authentication in medical device systems.* U.S. Patent 6,961,448.

Popp, M'raihi, & Hart. (2011). One-*time password.* U.S. Patent 8,087,074.

Rabkin, A. (2008). Personal knowledge questions for fallback authentication: Security questions in the era of Facebook. *Proceedings of the 4th symposium on Usable privacy and security,* 13-23. 10.1145/1408664.1408667

Rassan & Shaher. (2013). Securing mobile cloud using fingerprint authentication. *International Journal of Network Security & Its Applications, 5*(6), 41.

Rhee, K., Kwak, J., Kim, S., & Won, D. (2005). Challenge-response based RFID authentication protocol for distributed database environment. In *International Conference on Security in Pervasive Computing* (pp. 70-84). Springer. 10.1007/978-3-540-32004-3_9

Shanley, C. W., Jachimowicz, K., & Lebby, M. S. (1994). *Remote retinal scan identifier.* U.S. Patent 5,359,669.

Shaw, Holway, Alex, Nikolai, Joyce, Hilsenrath, & Speers. (2003). *Method and system for facilitating secure transactions.* U.S. Patent Application 10/032,535.

Shi, Y., Xie, W., Xu, G., Shi, R., Chen, E., Mao, Y., & Liu, F. (2003). The smart classroom: Merging technologies for seamless tele-education. *IEEE Pervasive Computing*, 2(2), 47–55. doi:10.1109/MPRV.2003.1203753

Soare, C. A. (2012). Internet banking two-factor authentication using smartphones. *Journal of Mobile. Embedded and Distributed Systems*, 4(1), 12–18.

Steeves & Snyder. (2007). *Secure online transactions using a CAPTCHA image as a watermark*. U.S. Patent 7,200,576.

Wayman, J., Jain, A., Maltoni, D., & Maio, D. (2005). *An introduction to biometric authentication systems. In Biometric Systems* (pp. 1–20). London: Springer.

Chapter 9

A Novel Bat Algorithm for Line-of-Sight Localization in Internet of Things and Wireless Sensor Network

Mihoubi Miloud

https://orcid.org/0000-0001-7892-0382
Djillali Liabes University, Algeria

Rahmoun Abdellatif
École Supérieure en Informatique, Algeria

Pascal Lorenz

https://orcid.org/0000-0003-3346-7216
University of Haute Alsace, France

ABSTRACT

WSNs have recently been extensively investigated due to their numerous applications where processes have to be spread over a large area. One of the important challenges in WSNs is secure node localization. Its main objective is to protect the circulated information in WSN for any attack with low energy. For this reason, recent approaches relying on swarm intelligence techniques are called and the node localization is seen as an optimization problem in a multi-dimensional space. In this chapter, the authors present an improvement to the original bat algorithm for information protecting during the localization task. Hence, the proposed approach computes iteratively the position of the nodes and studied the scalability of the algorithm on a large WSN with hundreds of sensors that shows pretty good performance. Moreover, the parameters are simulated in different scenarios of simulation. In addition, a comparative study is conducted to give more performance to the proposed algorithm.

DOI: 10.4018/978-1-5225-8100-0.ch009

INTRODUCTION

A wireless sensor network (WSN) is an ad-hoc network with a large number of nodes that are micro sensors capable of collecting and transmitting environmental data in an autonomous way. The nodes' positions are not necessarily predetermined, they can be randomly dispersed in a geographical area, called the "sensing area," corresponding to the area of interest for the phenomenon being captured. In WSN, a large number of nodes are deployed in the network, the information detected by the sensor node will be gathered and transmitted through multi-hop techniques to sink (Figure 1). I.e. each node sends the information to its neighbour (so one hop between two neighbours) until it reaches the last and sends it to sink.

Recently, WSN has become a very active research field, and the fruitful employment of WSN has opened various new research areas for application, such as climate prediction, analysis of sane, atmospheric pressure, and so on (Adnan et al., 2014), where several issues are addressed in the WSN, such as energy minimization, compression schemes, self-organizing network algorithms, routing, Protocols, security, cyber security (He et al., 2017) and quality of service management.

(Mao et al., 2007) Node localization is one of the important challenges of WSN, (Boukerche et al., 2007) it plays a vital role in several fields, such as coverage, deployment purposes, routing information, location service, target tracking, and mortar launching and monitoring in underwater WSN (see figure 2) .

The main objective of node localization is to estimate the sensor's location with initially unknown location information; in order, the process uses knowledge of the absolute positions of a few sensors and inter-sensor measurements, such as: distance and bearing measurements. The sensors with unknown location information are called non anchor nodes, while sensors with known location information are called anchor

Figure 1. WSN communication architecture

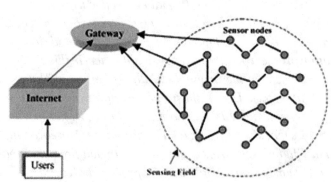

Figure 2. Monitoring in Underwater WSN

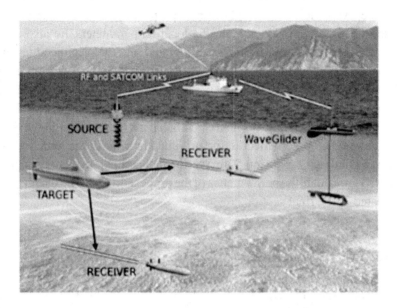

or beacons nodes. Self-localization capability is highly desirable in environmental monitoring applications such as intrusion detection, road traffic monitoring, health monitoring, and so on. One possible solution of node localization is to equip each node with global positioning system (GPS) devices, but this solution is not suitable for two reasons:

- The high cost of the device in terms of value, energy, computation power, and space.
- The poor precision of the service in special environments (indoors, underground, etc.).

The first tentative vs. node localization is proposed in Reference (Doherty et al., 2001), where convex optimization is proposed to localize the network nodes. Currently, localization requires each unknown node to have GPS installed but uses only a few anchor nodes, and it uses communication techniques proposed in References (Yick et al., 2008, Kulkarni et al., 2009) to localize the unknown nodes, where their coordinates will be estimated by the sensor network localization algorithm. In fact, the node localization problem has been considered a multidimensional optimization problem, where optimization algorithms are used to resolve this matter, and the recently invented Bat algorithm is proposed (Yang, 2011) as a metaheuristic algorithm. The frequency parameter is unchanged in the Bat algorithm during the shifting of the Bats; this factor makes the algorithm heavy, but the execution time is really considerable. An efficient Bat algorithm is proposed by updating the frequency parameter, moreover, the velocity and the location parameters are also modified, and the concept of the Doppler Effect is integrated into the original algorithm.

This paper is divided into 7 sections. Section 2 gives an overview about how to treat the node localization problem using a meta-heuristic algorithm, and related work is presented and analysed in this section. Section 3 presents a study for the Bat algorithm. An effective meta-heuristic algorithm is described in the Section 4 by integrating the Doppler Effect in the original algorithm. The node localization problem is formulated in Section 5. In Section 6, a comparative study is presented and analysed to evaluate the performance of the proposed algorithm. Our conclusions are elucidated in the final section.

RELATED WORK

Node localization has received much attention in the last few decades, and several surveys are detailed in References (Pal, 2010, Alrajeh et al., 2013, Vimee et al.,2014). The classification of node localization is divided after much consideration and according to several criteria:

- The mobility of the nodes (static vs. mobile, mixed).
- The execution environment (centralized vs. distributed, mixed).
- The distance estimation (range-free and range-based).

On the other hand, the localization scheme can be divided into 3 stages combined, as shown in Figure 2.

The principal task of the first stage is evaluating the distance between beacon nodes and target nodes, and several techniques such as Received Signal Strength Indication (RSSI),(Hoang et al., 2011) time of arrival (TOA) and time difference of arrival (TDOA),(Shen et al., 2010)and angle of arrival (AOA) (Yanping et al., 2008) are used to calculate the angle or distance.(Singh et al., 2011) Furthermore, the device to device technique has recently been proposed to link ad-hoc networks and Centralized networks, (Wang et al., 2017) and so on. Each of the estimation distance techniques possesses advantages and drawbacks, and the selection of the distance estimation method is considered an important factor, which greatly influences the performance of the system. In addition, each technique is specified according to the application field. The second stage of the localization scheme is the computation of position by exploiting the data obtained in the previous phase (distance estimation). It is possible to calculate the coordinates of the target nodes, and a surprising number of approaches are proposed in this phase; such methods include trilateration, Multilateration), triangulation, probabilistic approaches, bounding box, the central position, and fingerprinting. (Jiajia et al., 2015) Furthermore, it exists another work concentrates all information about distances computing that utilizes mathematical optimization methods to calculate the positions of the nodes. In the third stage of the process (localization scheme), in order to calculate the positions of the unknown nodes via beacon nodes, a combination operation is initiated between the previous two phases, There are considerable studies focused on the localization algorithm, Machine learning methods that were used to address common issues in wireless sensor networks (WSNs). The advantages and disadvantages of each proposed algorithm are evaluated against the corresponding problem (Alsheikh et al .,2014), The Bayesian approach are proposed as solution to deal with node localization (Morelande et al ., 2008), Neural Network (NN) is suggested as supervised model for node localization challenge (Shareef et al ., 2008),the computational resource and memory resources are evaluated under several kind of NN .

Recently, metaheuristics methods have been considered important to solve the optimization problem, whereas node localization is treated as a multidimensional optimization problem, and a substantial number of methods that use population-based stochastic techniques are proposed.

In 2008, particle swarm optimization (PSO) was proposed as the first attempt to solve the node localization issue, (Gopakumar et al., 2008) and a few anchor nodes were used to locate the remaining unknown nodes in the square network. The proposed method was based on assembled information by anchor nodes to locate the rest of the sensors. The basic idea of this approach was that all unknown nodes send their calculated coordinates and estimated distance to the base station and the base station runs a PSO to minimize the localization error defined.

In 2011, the Bee optimization algorithm (BOA) was proposed for node localization (Moussa et al., 2010), where TOA and RSSI are utilized in the estimation phase. On the other hand, 2 deployment models are executed in the simulation network, the first one is the best deployment, in which the target nodes are surrounded by 4 (placed in the corner) beacon nodes in different directions, and the second is in a worse environment where the beacon nodes are deployed in the medium oriented in a single direction relative to 1 unknown node. The obtained result proved that BOA was efficient and has remarkable performance. Furthermore, a comparative study was conducted between the proposed approach and the Cramer-Rao Bound (CRB), and the final result demonstrated that the accuracy rate of the proposed approach based on TOA is better than the RSSI measurement.

In 2014, differential evolution (DE) was proposed for the node localization challenge, (Harikrishnan et al., 2014) in order to minimize the localization error, each node in the network runs DE to determine its coordinates. The approach has proven that converges after a few iterations (22) allow the conservation of energy and the cost of the nodes.

Flower pollination (FP) was another meta-heuristic technique that was proposed in 2015, (Goyal et al., 2015) for the node localization problem. The fundamental idea is to evaluate the coordinates of the best nodes by bringing these neighbours closer by one hop. A comparative work is reviewed in this paper, and the simulation results obtained (the coordinates of calculated nodes) by FP were compared with the various improvements of the PSO algorithm. In addition, several topologies of network are exercised, and the simulation results show that the localization rate obtained by the FP algorithm is higher than the various improvements of the PSO algorithm.

The Bat algorithm is a recent meta-heuristic proposed in Reference 7. It is a method developed for global numerical optimization, and the main idea is inspired by the echolocation behaviour of bats by changing pulse rates of emission and loudness.

The Bat algorithm has been proposed for the node localization problem in Reference (Goyal et al., 2013), and the proposed approach has proven its efficiency. The obtained results show that the localization rate by the proposed algorithm is very important. Moreover, the Bat algorithm can arrive at an estimated rate higher than the intake by BBO. The Bat algorithm has demonstrated it capacity and adaptability for the node localization problem, but fares poorly in the exploration phase. The Bat algorithm is powerful at exploitation but demonstrates some insufficiency in the exploration phase; in that it cannot explore all the bats in the search area.

In 2016, the Dragonfly algorithm (DA) was proposed for the node localization problem in Reference (Philip et al., 2016). A maximum of 10 unknown nodes are deployed in the square network where the anchor nodes are placed in the corner; several parameters are simulated to prove the development and resistance of the proposed algorithm. A comparative study was conducted with PSO, and the simulation results showed that the localization error obtained by DA is less than PSO .

Grey Wolf Optimization (GWO) is another metaheuristic proposed to deal with for node localization problem (Rajakumar et al., 2017), the obtained result demonstrate that GWO converge rapidly and error localization is minimal, the simulations results are compared with MBA (modified Bat algorithm and PSO), GWO proved that has faster convergence rate and estimating the location of the node so the computation time is minimal as advantages and furthermore GWO has easy steps and few parameters . a comparative study is introduced between the three most popular metaheuristic (Bat Algorithm, PSO and DE) in (Mihoubi et al., 2017), whereas two criteria parameter are evaluated: localization error and localization time, the obtained result shown that Bat algorithm gives more efficiency and accuracy rate is better than the two other.

An effective metaheuristic (Dopeffbat) is proposed to deal with localization problem (Miloud et al., 2017), the simulation with Dopeffbat get remarkable performance result and proves is efficiency to other approaches.

Moth Flame Optimization Algorithm (MFOA) is the latest approach proposed to deal with node localization in distributed paradigm of WSN (Miloud et al., 2019), MFOA it is show are combination between PSO and genetic algorithm (GA), furthermore, the moth fly smartly through several trajectory as spiral trajectory, the simulations result prove that MFOA is very efficient and improve the performance of the network .

Important Motivation of Cyber Security in WSN

WSNs have been enormously used in several domain surveillance and security functions. As example, sensor networks have been deployed to support surveillance capabilities such as threat-presence detection within security-sensitive and hostile regions such as a militarized area, border protection, etc. To support monitoring, a WSN has been proposed in (S. Bitam . 2016) to detect and determine the direction of movement of intruding personnel and vehicles in the sensitive region.

Deployed sensors cooperate with each other to detect an imminent approaching mobile threat and are able to auto-organize to provide a relevant, timely, and concise net-centric view of the monitoring field. This information aides to improve decision abilities for intervenes intelligence, monitoring.

To improve this, the nodes should be capable to detect the event and gathered the information and forward to sink or head of the region. In such an environment, sensor nodes may be protected to any attack of the interior or from the outside.

Medical monitoring can also be cited as a healthcare service provided by wearable and implantable body sensors connected in the well-known body sensor network (BSN). A typical BSN is composed of a number of miniature, lightweight, low-power sensing devices and wireless transceivers. These sensor networks are used to capture large amounts of data containing information about the patient's health status, which is then stored in some database.

The collected information in Healthcare can contain blood pressure, heart rate, distance travelled through walking/running, playing activities, and surroundings (e.g. room temperature).

This collected information helps to monitoring the patient from any place and can intervenes at any moment immediately as monitoring of chronic disease, elderly people, postoperative rehabilitation patients, and persons with special disabilities (Sharma et al.,2018).

However, in this case a BSN may be exposed to a malicious party who can exploit various serious security threats to compromise the healthcare service and prevent patients from reaching available healthcare facilities

BAT ALGORITHM (BAT)

Bat Algorithm is novel Metaheuristic population based algorithm Developed (Yang, 2011), it inspired by the echolocation behaviour of fascinating animals named Bats, the micro Bats utilize an echolocation to finds its prey, locate their roosting crevices in the dark and avoid obstacles, this micro bat has the capacity to locate positions of prey by sending high and short audio signals, by collision and via the echo of signal sender, The advanced echolocation capability of Bats makes them fascinating and work perfectly in darkness environment .Substantially, the main five steps of Bat algorithm are:

Step 1: Initialization of Bat Population

Firstly, N bats are randomly spread over the search space as they have no idea where the prey is located thus population is randomly generated for each dimension d with default values for frequency, velocity, pulse emission rate and loudness.

$$x(i,j) = LL(j) + rand(0,1).UL(j) - LL(j) \tag{1}$$

whither $\in [1..N]$ is a random vector drawn from a uniform distribution, i = 1, 2 … N, j = 1, 2… d, the value of UL (upper limit) and LL (lower limit) depending on the network area (X_i, Y_i).

Step 2: Movement of Virtual Bats (Update Frequency, Velocity and Position)

In the iterations of Bat, each bat in the population emits sound pulses of random frequency and this frequency controls the speed and the updated positions (new solution) and new speed at time step t are regularized by:

$$f_i = f_{min} + \left(f_{max} - f_{min}\right).rand \tag{2}$$

$$v_i^{t+1} = v_i^t + \left(x_i^t - x^*\right).f_i \tag{3}$$

$$x_i^{t+1} = x_i^t + v_i^{t+1} \tag{4}$$

where $f_i \in \left[f_{min}, f_{max}\right]$ is frequency, $f_i, f_i \in \left[f_{min}, f_{max}\right] \cdot x_i^t$ is position, v_i^t is velocity, $rand \in \left[0, 1\right]$ is a random vector drawn from a uniform distribution, x_* is the current global best solution.

Step 3: Local Search

For the local search region, only one solution is chosen among the actual best candidates, browsing the Bat candidates groups a new solution is created locally based on random walk according to the following equation:

$$if(rand\left(0, 1\right) > r_i \tag{5}$$

$$x_{new} = x_{old} + \varepsilon + A^t \tag{6}$$

where $\varepsilon \in \left[-1, 1\right]$ is a random vector drawn from a uniform distribution, A^t is the medium loudness of the bats at step t, $r_i \in \left[0, 1\right]$ is the pulse rate. If the best value obtained by the total N bats is higher than the precedent $f(x_*)$, at this time the global best solution x_* can be updated.

Step 4: Updating Loudness and Pulse Emission Rate

If the new solutions are updated so the loudness and emission rates decrease and increase respectively, this means that these bats are headed for their prey. This relationship can be explained as follows:

$$if(rand\left(0, 1\right) < A_i^t \,\&\, \&f(x_i) < f(x) \tag{7}$$

$$f(x) = f(x_i) \tag{8}$$

$$A_i^{t+1} = \alpha . A_i^t \tag{9}$$

$$r_i^{t+1} = r_i^0 . (1 - e^{-yt}) \tag{10}$$

where $\alpha \in [0,1]$ and $y(y > 0)$ and y (y > 0) are constants, r_i^0 $\varepsilon \in [-1,1]$ are a random value, $r_i^0 \in [0,1]$, ($A_i^t \in [1,2]$ at t = 0) .

Step 5: Find the Current Best Solution

Find the current best bat by comparing last best fitness value with improved bat fitness if it's better so update the best bat solution. When the last iteration is reached, the current best bat x_* gets improved and acts as the solution to the problem.

Improved Bat Algorithm (IBat)

An improved Bat algorithm (IBAT) was formulated based on the changing frequency of the periodic event when an observer moves relative to its source. The Doppler Effect is the modification of the wavelength of a source due to the relative movement between the receiver and the sources. The relative movement that affects the observed frequency is the movement in the line of sight (LOS) between the transmitter and the receptor.

The bats have better vision than humans; they have a high-frequency system through which they can easily locate and hunt prey even in the dark. This system is called echolocation. The bat sends a pulsation when it moves and listens to the returning echoes to obtain a sonic map of its surrounding environment. The Bat can easily determine the distance and the time necessary for the sound it makes to echo back.24 the bats fly randomly in the environment and update their frequency Doppler to save the echoes cantered in the restrained frequency measurement area. They possess a very sensitive hearing ability, and this characteristic is because of the Doppler shift compensation. The advantage of the Doppler shift is the ability to hear the beating of wings as slight deviations of frequency within the bat's hearing range.

Two Doppler shifts occur to calculate the frequency heard via echolocation. Suppose ft/vt and fr/vr are the frequencies/speeds of the transmitter and the receiver, respectively. The original frequency has a relation with the wavelength, where $f0 = C/f0$, $f0$ is the wavelength of the source, and C is the wave speed in the medium (in air, $c = 340$ m/s)

If the source is displacing forward from the observer, the velocity of flying is negative, so: (11)

If the source is moving away from the observer, the velocity of flying is positive; so:

$$f' = f_0 \left[\frac{C - v_r}{C} \right] \tag{12}$$

The 2 equations (11) and (12) are usually combined and expressed as:

$$f' = f_0 \left[\frac{C \pm v_r}{C} \right] \tag{13}$$

When the source is displacing toward the receptor, the equation becomes:

$$f' = f_0 \left[\frac{C}{C - v_s} \right] \tag{14}$$

If the source evolves, the new frequency will be updated to:

$$f' = f_0 \left[\frac{C}{C + v_s} \right] \tag{15}$$

By combining equations (14) and (15), the novel frequency would be:

$$f' = f_0 \left[\frac{C}{C \mp v_s} \right] \tag{16}$$

The plus/minus symbol is inverted because the sign on top is to be used for the relative motion of the source toward the receiver.

The deducted Doppler Effect equation is expressed as:

$$f_r = f_t \left[\frac{C \pm v_r}{C \mp v_r} \right] \tag{17}$$

The new frequency equation is integrated with the original Bat algorithm:

$$f_i^{t+1} = f_i^t \left[\frac{C \pm v_i}{C \mp v_i} \right] \tag{18}$$

In addition, we added new parameters ($w1$, $w2$) to improve the performance of the algorithm and regularized the result.

The parameters are capable of balancing the proportional relationship between the global convergence ability and the local convergence in general, ($w1$, $w2$) C (0.7, 1).

$$v_i^{t+1} = w_1 * v_i^t + \left(x_i^{t+1} - x_* \right) \times f_i^{t+1} \tag{19}$$

$$x_i^{t+1} = w_2 * x_i^t + v_i^{t+1} \tag{20}$$

NODE LOCALIZATION PROBLEM AS OPTIMIZATION PROBLEM

The node localization in WSN consists of calculating the coordinates of the maximum unknown nodes (target nodes) based on the minimum anchor nodes with 1 hop-based range in the distributed network.

The main steps of the node localization scheme are:

1. N beacons nodes and M unknown nodes are deployed randomly in the network, associated with the same communication range as N and M.
2. The beacon nodes calculate the positions and transmit their coordinates based on RSSI for the position estimation of unknown nodes.
3. Each node that is surrounded by 3 or more anchors is considered a localized node.
4. Each node that becomes localized calculates its distance di from each of its neighbouring beacons by trilateration (N = 3) or Multilateration (N > 3). A real distance between the target node (x, y) and i-th anchor (xi, yi) is described as:

$$d_i = \sqrt{\left(x - x_i \right)^2 + \left(y - y_i \right)^2} \tag{21}$$

The distance measurements are usually affected by multi-path and obstacles blocking due to environment consideration, which can represented as noise n_i, the measured distance $\hat{d}_i$ between the target node (x, y) and ith anchor (x_i, y_i) is modeled by equation:

$$\hat{d}_i = \left[d_i + n_i \right] \tag{22}$$

5. Each node localized runs improved Bat algorithm independently to locate itself by calculates its coordinates (x, y). The objective function (mean square error between the actual node coordinates and estimated distances), is calculated by:

$$f(x, y) = \frac{1}{N} \sum_{i=1}^{N} \sqrt{(x - x_i)^2 + (y - y_i)^2} - \hat{d}_i \tag{23}$$

where N ≥ 3 (2 D the location of a node requires at least 3 anchor nodes) is the number of beacon nodes with a transmission range

6. The nodes which get established at the end of iteration will be added to the sets of nodes and will be play as anchors nodes in the next iteration.
7. The steps 2 to 6 are repeated until stopping condition is reached (no unknown node can be localized or

Number iteration terminated or tolerance error) .the Mean Localization Error (MLE) is calculated as the mean square of Euclidean distance so of estimated node coordinates (x_i, y_i) and the real node coordinates (X_i, Y_i) for i = 1, 2...ML determined for improved Bat algorithm, as shown below:

$$E_L = \frac{1}{M_L} \sum_{i=1}^{L} \sqrt{(x_i - X_i)^2 + (y_i - Y_i)^2} \tag{24}$$

where "L=M-N ", M is the number of localized node.

Simulation Experiments and Performance Evaluation

The performance evaluation of the improved Bat algorithm has been evaluated by studying several parameters, such as population size, iteration number, communication range, anchor density, in addition, a comparative study with original Bat algorithm and PSO algorithm has been evaluated by introducing several criteria using the same network deployment.

1. **The Fundamental Metric:** Two major metrics (Abdi et al., 2014) are proposed for performance evaluation:
    a. **Average Localization Errors (ALE):** this is equal to average distance between the estimated location (x_i, y_i) and the actual node coordinates (X_i, Y_i).

$$E_L = \frac{1}{M_L} \sum_{i=1}^{L} \sqrt{\left(x_i - X_i\right)^2 + \left(y_i - Y_i\right)^2}$$
(24)

    b. **Average Execution Time (AET):** which is the average time required for localization of all sensor nodes

$$Average\,excution\,time = \frac{\sum_{i=1}^{n} excution\,time}{n\left(number\,of\,sensor\,node\right)}$$
(25)

2. **Simulation Setup:** The simulations are performed using MATLAB environment on a laptop of 4 GB memory and i5 -2.2 GHz CPU to evaluate the performance of proposed algorithm.

In our simulation experiments 100 sensor nodes (unknowns and anchors) are randomly deployed in a 50 m ×50 m square simulation area (figure 3), where the anchor nodes are 40% of the total sensors nodes, the communication range of the anchor is 15 m.

The figure 3 shows the nodes deployments in square area, the green nodes represent the anchors nodes and the blue star nodes represent the unknown nodes.

The neighbour relationship diagram according to previous deployment are displayed in figure 4, wherein the communication range is 15 m, the neighbour relationship diagram explain that the unknown nodes are in the audience of the anchors nodes or not .

The exact position of the target nodes estimated by improved Bat is presented in figure 5, the triangle shape represents the resolved nodes .

3. **Localization Results Under Different Population Size:** This part is devoted mainly to study the parameters of simulation, we analyse the plausible

Figure 3. Nodes deployment in network

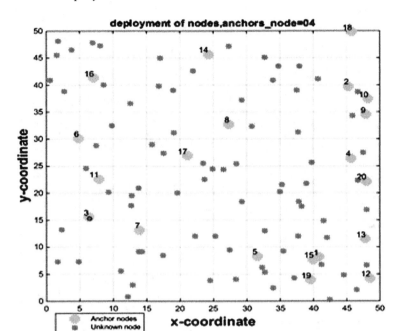

influences on the rate of localization, and the parameters are the population size, communication range, and anchor density.

The population size has an essential task for the convergence of the algorithm and mainly in the phase of exploration of the neighbours.

The simulation results schematized in figure 6 shows that the algorithm converges rapidly when the population rate is large, This characteristic returns to the advantage of the algorithm in the exploration phase where

The generation process of candidate solution around the best solution increases exploitation capability, on the other hand, from the figure 7, we found that localization time has an inverse relationship with the localization error, this event is interpreted that during the exploitation phase of the algorithm each bat exploits all the neighbors to have the best coordinates.

4. **Localization Results Under Different Communication Range:**
    Communication range is another important parameter determining localization performance and energy consumption of sensor nodes. Figure 8 explains the impact of communication range of localization error under different node

Figure 4. Communication connectivity between sensor nodes

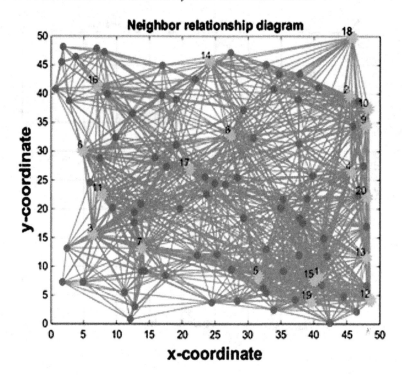

density, the parameters are the same for the before network where the total sensors nodes is 40, 20% for anchors nodes.

In the other side, from figure 9 when the communication range is greater than 20 the localization error begins to decrease this is because there is more anchor information available for computing the location of unknown nodes.

Whereas when communication range reaches certain value so the localization error slowly decreases localization time versus communication range is presented in figure 9 we found that the proposed algorithm has convergence speed, this is due to the fact that when the communication range is short the anchor nodes localize the unknown node rapidly but the error is large, while the when communication range increase the localization time will be long, note that the noise of the environment make it influence to settled the unknown node and over localization error, we conclude that the localization time is very good and converge with the communication range this allows keeping the energy consumption.

5. **Localization Results Under Different Anchor Density:** The density of the anchor nodes is essential parameter which influences over localization

Figure 5. Node localization by improved Bat (100 nodes)

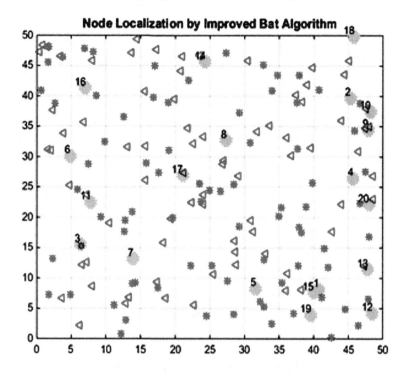

Figure 6. Localization Error versus Population size

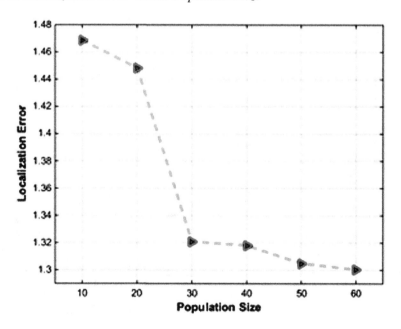

Figure 7. Localization Time versus Population size

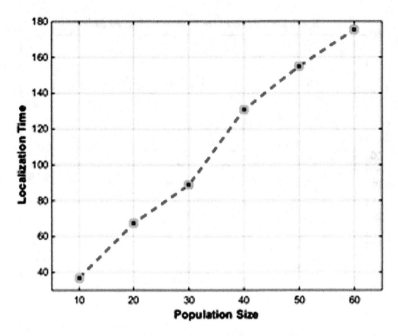

Figure 8. Localization Error versus Communication Range

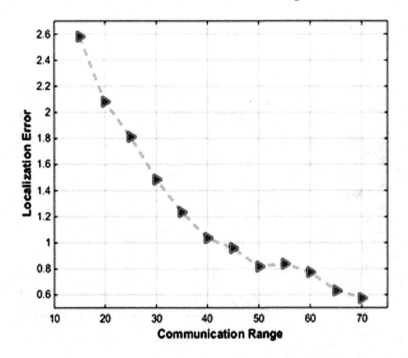

Figure 9. Localization Time versus Communication Range

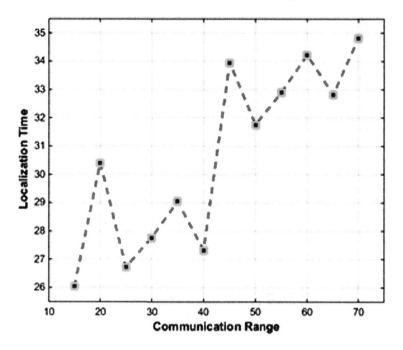

performance and cost for WSN. In this subsection, to determinate the influence of this parameter, several experiments were carried out, the anchor ratio is set as follow: 10%, 20%, 30%, 40%, 50%, 60%, 70% and 80% of all sensor nodes, the communication range of each sensor node is 20 m, figure 10 presents the variation of localization error under the impact of anchor density, we observe that the first value of anchor density is 0.1 of nodes totality (100) equal to 4 anchor nodes, this number is enough to make localization in two dimension where the minimum anchor nodes are 3. We observed that the average localization error is decreased gradually with respect to the increasing of anchor density, when node density increases, the network connectivity between sensor nodes becomes high and the number of anchor nodes increases.

Effectively, on the other side, from figure 11 we observe that localization time decrease with each increase of anchor nodes, this is due the fact that at each iteration new localized nodes became an anchors node that decreases the number of unknown nodes.

Figure 10. Localization Error Anchor Ratio

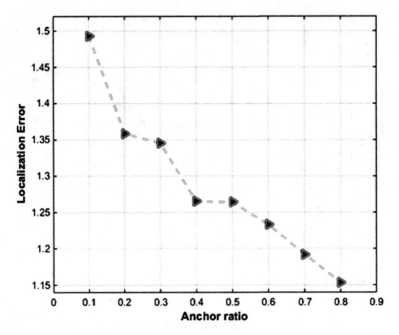

Figure 11. Localization Time versus Anchor Ratio

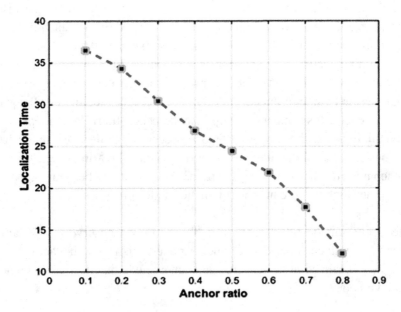

6. **Comparative Study of Improved Bat With Original Bat Algorithm, PSO on the Basis of Localization Error:** This section evaluates the performance of the proposed approach "improved Bat algorithm," where a comparative study was carried out with the original Bat algorithm and PSO

The total number of sensor nodes is 100 nodes, the deployment of sensors nodes is random in 50 m × 50 m network area, 96 unknown nodes and to avoid the collinear problem 4 anchors nodes (red colour) are deployed in the four corner of network area (Patwari et al., 2005), the coordinates of anchors nodes are (0,0), (0,100), (100,0), (100,100), In addition, it is assumed that the communication between the nodes is feasible if only if the nodes are in the range of network, the communication connectivity is schematized in Figure 11, communication range for all sensor nodes is set to 50 .

A comparative study is conducted for evaluating the performance of metaheuristic algorithm, several scenarios are simulated, analysed and evaluated in detail in (David et al ., 2008), although the results obtained and analysed in detail in (Jonesa et al., 2006) proved that Bat algorithm is better and advantageous against PSO.

Figure 12. Deployment of network

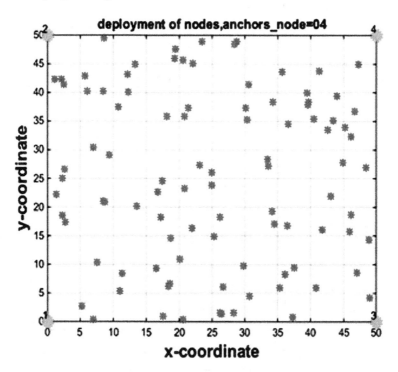

The initials parameters of PSO are initialized according to as cited in (Gopakumar et al., 2008, Singh et al., 2011). The parameters for original Bat algorithm are figured in table 1, the adding parameters for the proposed algorithm (improved bat algorithm) are described in table 2.

WSN (Wireless Sensor Network) is an ad hoc network with a large number of nodes that are micro-sensors capable of collecting and transmitting environmental data in an autonomous way. The nodes position is not

Table 3 illustrates the Mean Localization Error (MLE), localization time obtained by the three algorithms versus the iteration number during localization scheme, from table 3.a we saw that as the mean localization error is increasing rapidly by the proposed algorithm than the original Bat and PSO, this is due the variation of frequency of each Bat, when bat search her neighbours by sending pulsation.

When got echo it will adapt his frequency more appropriate with velocity, this modification optimizes the energy consumption of the network. On the other hand, the localization time (table 3.b) of the proposed algorithm is small and converge very quickly, in addition, more velocity increase more the localization time decrease and occupation of node decrease.

Table 1. The parameter value for Bat

Parameters	Parameters Value
Population size	10
Number of generations	100
Loudness	0.99 - 2
Pulse rate	1
Minimum Frequency	0
Maximum Frequency	1
Dimension	2
Alpha	0.9
Sigma	0.9

Table 2. The parameter value for Improved Bat IBAT

Parameters	Parameters Value
wMax	0.9
wMin	0.5
Vmax	10

We remark that at iteration number 15 the localization error stable, this is due the initial parameters of the both algorithms, we concluded that the accuracy rate of improved Bat algorithm is better than PSO and original Bat algorithm.

CONCLUSION

The Bat algorithm is the most popular metaheuristic that has proven its efficiency and profitability, but Bat prone to two drawbacks, the convergence rate and the localization error. In order to overcome the two matters, an improvement is proposed to perform the algorithm by updating the frequency, changing frequency make algorithm converges rapidly during the bat and minimizes the localization error. The results obtained demonstrated that the modification tackled on the original

Table 3. Comparative study of the three algorithms IBAT

Iteration Number	Range	Mean Localization Error		
		Improved Bat	Bat	PSO
2	50	1.0490	1.0249	1.0851
5	//	0.8365	0.9817	1.0692
10	//	0.8270	0.9104	1.0560
15	//	0.8111	0.8201	1.0561
20	//	0.7967	0.8015	1.0541
25	//	0.7833	0.7991	1.0535
30	//	0.7640	0.7909	1.0530

Table 4. Comparative study of the three algorithms IBAT

Iteration Number	Range	Localization Time		
		Improved Bat	Bat	PSO
2	50	11.9368	13.7587	10.5954
5	//	19.2556	22.2356	19.9842
10	//	31.4827	42.2361	37.7122
15	//	43.9885	54.4526	49.6678
20	//	65.9765	77.7812	68.1125
25	//	88.1953	103.9467	91.3470
30	//	102.0056	125.4635	110.3139

algorithm is performing and the criteria studied are enhanced where localization error and localization time are better than the other two algorithms.

We focused on node localisation and looking for cyber security the circulation of the sensors nodes information, which is an integral part of many approaches. In reviewing various swarm intelligent approaches to improve the cyber security of localization, we found that there is a need to address several of the drawbacks of recently proposed and integrating machine learning technique as (Conventional neural network, Recurrent Neural Network).

REFERENCES

Abdi, F., & Haghighat, A. T. (2014). A hybrid rssi based localization algorithm dor wsn using a mobile anchor node. *International Conference on Computing, Communication and Networking Technologies*.

Adnan, M. A., Razzaque, M. A., Ahmed, I., & Isnin, I. F. (2014). Bio-Mimic optimization strategies in wireless sensor networks: A survey. *Sensors (Basel)*, *141*, 299–345. PMID:24368702

Alrajeh, N. A., Bashir, M., & Shams, B. (2013). Localization techniques in wireless sensor networks". *International Journal of Distributed Sensor Networks*, *96*, 1–9.

Alsheikh, M. A., Lin, S., Niyato, D., & Tan, H. (2014). Machine Learning in Wireless Sensor Networks: Algorithms, Strategies, and Applications. IEEE Communications Surveys & Tutorials, 16(4), 1996-2018. doi:10.1109/COMST.2014.2320099

Bitam, S., Zeadally, S., & Mellouk, A. (2016). Bio-inspired cybersecurity for wireless sensor networks. *IEEE Communications Magazine*, *54*(6), 68–74. doi:10.1109/MCOM.2016.7497769

Boukerche, A., Oliveira, H., Nakamura, E., & Loureiro, A. (2007). Localization systems for wireless sensor networks. *IEEE Wireless Communications*, *14*(6), 6–12. doi:10.1109/MWC.2007.4407221

David, S., Martikainen, J., Dudoit, S., & Ovaska, S. J. (2008). A general framework for statistical performance comparison of evolutionary computation algorithms. *Information Sciences*, *17814*, 2870–2879.

Doherty, L., Pister, K. S. J., & Ghaoui, L. E. (2001). Convex position estimation in wireless sensor networks. *Twentieth Annual Joint Conference of the IEEE Computer and Communications Societies Proceedings, 3*, 1655–1663.

Gopakumar, A., & Jacob, L. (2008). Localization in wireless sensor networks using particle swarm optimization. *International Conference on Wireless, Mobile and Multimedia Networks*, 227–230.

Goyal, S., & Pattenh, M. (2015). Flower pollination algorithm based localization of wireless sensor network. *2nd International Conference on Recent Advances in Engineering and Computational Sciences*, Chandigarh, India. 10.1109/RAECS.2015.7453299

Goyal, S., & Patterh, M. S. (2013). Wireless sensor network localization based on BAT algorithm. *Int J Emerg Technol Comput Appl Sci*, *5*, 507–512.

Harikrishnan, R., Jawahar, S. K., & Sridevi, P. (2014). Differential evolution approach for localization in wireless sensor networks. *IEEE International Conference on Computational Intelligence and Computing Research*. 10.1109/ICCIC.2014.7238536

He, Chan, & Guizani. (2017). Cyber Security Analysis and Protection of Wireless Sensor Networks for Smart Grid Monitoring. *IEEE Wireless Communications, 24*(6), 98-103. doi:10.1109/MWC.2017.1600283WC

Hoang, Q. T., Le, T. N., & Shin, Y. (2011). An RSS comparison based Localization in wireless sensor networks. *8th workshop on Positioning Navigation and communication WPNC*, 116–121. 10.1109/WPNC.2011.5961026

Jiajia, L., Nei, K., Jianfeng, M., & Naoto, K. (2015). Device-to-device vommunication in LTE-advanced networks: A survey. *IEEE Communications Surveys and Tutorials*, *17*(4), 1923–1940. doi:10.1109/COMST.2014.2375934

Jonesa & Teeling. (2006). The evolution of echolocation in bats. *Trends in Ecology and Evolution, 213*, 149–156.

Kulkarni, R., Venayagamoorthy, G., & Cheng, M. (2009). Bio-inspired node localization in wireless sensor networks. *International Conference on Systems, Man and Cybernetics*, 205–210. 10.1109/ICSMC.2009.5346107

Mao, G., Fidan, B., & Anderson, B. D. (2007). Wireless sensor network localization techniques ". *Computer Networks*, *51*(10), 2529–2553. doi:10.1016/j.comnet.2006.11.018

Mihoubi, M., Rahmoun, A., & Lorenz, P. (2017). Metaheuristic RSSI based for node localization in distributed wireless sensor network. *Global Information Infrastructure and Networking Symposium (GIIS)*, 64-70. 10.1109/GIIS.2017.8169811

Miloud, M., Abdellatif, R., & Lorenz, P. (2019). Moth Flame Optimization Algorithm Range-Based for Node Localization Challenge in Decentralized Wireless Sensor Network. *International Journal of Distributed Systems and Technologies, 10*(1), 82–109. doi:10.4018/IJDST.2019010106

Miloud, M., Rahmoun, A., Lorenz, P., & Lasla, N. (2017). An effective Bat algorithm for node localization in distributed wireless sensor network. *Security and Privacy.* . doi:10.1002py2.7

Morelande, M., Moran, B., & Brazil, M. (2008). Bayesian node localisation in wireless sensor networks. *IEEE International Conference on Acoustics, Speech and Signal Processing*, 2545–2548. 10.1109/ICASSP.2008.4518167

Moussa, A., & El-sheimy, N. (2010). Localization of wireless sensor network using Bees optimization algorithm. *IEEE International Symposium on Signal Processing and Information Technology*, 478–481. 10.1109/ISSPIT.2010.5711760

Pal, A. (2010). Localization algorithms in wireless sensor networks: Current approaches and future challenges. *Netw Protoc Algorithm, 21*, 45–73.

Patwari, N., Ash, J. N., Kyperountas, S., Hero, A. O., Moses, R. L., & Correal, N. S. (2005). Locating the nodes: Cooperative localization in wireless sensor networks. *IEEE Signal Processing Magazine, 224*(4), 54–69. doi:10.1109/MSP.2005.1458287

Philip, T. D., & Soo, Y. S. (2016). Range Based Wireless Node Localization Using Dragonfly Algorithm. *Eighth International Conference on Ubiquitous and Future Networks (ICUFN)*. DOI: 10.1109/ICUFN.2016.7536950

Rajakumar, Amudhavel, Dhavachelvan, & Vengattaraman. (2017). GWO-LPWSN: Grey Wolf Optimization Algorithm for Node Localization Problem in Wireless Sensor Networks. *Journal of Computer Networks and Communications.* . doi:10.1155/2017/7348141

Shareef, A., Zhu, Y., & Musavi, M. (2008). Localization using neural networks in wireless sensor networks. *Proceedings of the 1st International Conference on Mobile Wireless Middleware, Operating Systems, and Applications*, 1–7. 10.4108/ICST.MOBILWARE2008.2901

Sharma & Bhatt. (2018). Privacy Preservation in WSN for Healthcare Application. *Procedia Computer Science, 132*, 1243 – 1252. . doi:10.1016/j.procs.2018.05.040

Shen, G., Zetik, R., Yan, H., Hirsch, O., & Thomä, S. (2011). Time of arrival estimation for range based localization in UWB sensor networks. *Proceedings of IEEE International Conference on Ultra -Wideband, 2*, 1– 4.

Singh, P. K., Tripathi, B. T., & Singh, N. P. (2011). Node localization in wirless sensor network. *Int J Comput Sci Inform Tech*, *26*, 2568–2572.

Vimee, W., & Simarpreet, K. (2014). Survey of different localization techniques of wireless sensor network. *IJEEE*, *14*, 759–767.

Wang, X., Lingjun, G., Shiwen, M., & Santosh, P. (2017). CSI-based fingerprinting for indoor localization: A deep learning approach. *IEEE Trans Veh Tech*, *66*, 763–776.

Yang, X. S. (2011). Bat algorithm for multiobjective optimization. *International Journal of Bio-inspired Computation*, *35*(5), 267–274. doi:10.1504/IJBIC.2011.042259

Yanping, Z., Daqing, H., & Aimin, J. (2008). Network localization using angle of arrival. *International Conference on Electro Information Technology*, 205–210.

Yick, J., Mukherjee, B., & Ghosal, D. (2008). Wireless sensor network survey. *Computer Networks*, *52*, 2292–2330.

Chapter 10

Ensemble Learning Mechanisms for Threat Detection:
A Survey

Rajakumar Arul
https://orcid.org/0000-0002-1385-7965
Amrita Vishwa Vidyapeetham, India

Rajalakshmi Shenbaga Moorthy
St. Joseph's Institute of Technology, India

Ali Kashif Bashir
Manchester Metropolitan University, UK

ABSTRACT

Technology evolution in the network security space has been through many dramatic changes recently. Enhancements in the field of telecommunication systems invite fruitful security solutions to address various threats that arise due to the exponential growth in the number of users. It's crucial for upgrading the entire infrastructure to safeguard the system from specific threats. So, there is a huge demand for the learning mechanism to realize the behavior of attacks. Recent upcoming technologies like machine learning and deep learning can support in the process of learning the behavior of all types of attacks irrespective of their deployment criteria. In this chapter, the analysis of various machine learning algorithms with respect to a few scenarios that can be adopted for the benefits of improving the security standard of the network. This chapter briefly discusses various know attacks and their classification and how machine learning algorithms can be involved to overcome the popular attacks. Also, various intrusion detection and prevention schemes were discussed in detail.

DOI: 10.4018/978-1-5225-8100-0.ch010

INTRODUCTION

The synonym of cyber is given as "through use of a computer" also called cyberspace. The term security refers to safe. These two words joined to form cyber security which means making cyberspace safe from threats. Over the past decade, the adventurous growth of technology imposes severe challenges on security. The technology keeps on growing in three dimensions viz. computation, storage, and connectivity. When the technology grows the threats associated with it also grows, i.e., the growth of technology is directly proportional to the threats possible. Thus Security is the prime concern for Information and Communication Technology (ICT). The various ICT devices such as smartphones, sensors, actuators, RFID devices are interdependent and communicate with one another, which increases network traffic, is subjected to various security attacks. The mechanism of giving protection to ICT Systems is termed to be cybersecurity. Cybersecurity is the act of i) securing a computer, and its associated hardware, software, and data stored in it ii) securing computer networks, its associated hardware, software and data transferred across it. Thus to provide cybersecurity it is essential to understand about attacks, and the mechanisms to secure against those attacks. A good security system must protect the following objectives: i) confidentiality, refers to unauthorized access of information ii) Integrity, refers to the unauthorized modification of information iii) Availability, refers to the denial of service of prevention of information from authorized access. The attacks in the networks are broadly classified into two categories: i) Passive Attack, which is silently listening communication ii) Active Attack which is modifying the data transferred across the network. Security is the process, and it is not an end state. Several mechanisms are used in day to day life to impose maximum security.

In this pervasive world, the technology changes our lifestyle, the way we did business, and the way we had communication with friends, relatives or neighbors. It was all made possible with one word "Internet." When everyone is using the internet, a huge amount of data gets generated, transferred across networks and stored in the system. Thus the growth of Internet sophisticates people's life and at the same time it paves the growth of Cyber criminals. Cybercriminals are those who steal one's personal information stored on the computer. Even though there are security mechanisms, the criminals find a hole to come out of it. Thus there is always intense violence between cybercriminals and cyber security providers. The conventional security mechanisms always focus on protection from attack rather than detection of an attack. Traditionally network is protected using a firewall, and the computer system can be protected using antivirus software. Traditional threat protection mechanism depends on signature and a pattern. When a threat to the system ever seen before arises, will the tool able to protect the system is a miserable question?

A good threat detection algorithm should efficiently monitor the behavior at the network level, host level, and user level. If any abnormal behavior occurs, which the system had ever seen then the algorithm should report it as abnormal behavior.

New technology like IoT, which interconnects all devices, gives a huge attack space for cybercriminals. Such attacks are generated at a fast rate, persistent, voluminous and continually evolving over time. The increase of threats and to increase the speed of processing the vast amount of data, machine learning can be used. The use of machine learning not only improves the speed but also provide better insights. Cyber attack detection algorithm has to detect the attacks accurately by acquiring and analyzing the past data. An excellent Cyber attack detection algorithm should improve the accuracy.

Some of the cyber attack detection algorithms include Signature Based Methods, Anomaly-based methods. The Signature based methods detect attacks that follow a well-known pattern (signatures). The problem with this approach is that it mainly identifies the existing attack and it is not possible to determine future attacks The Anomaly-based Method search for unusual behavior in a system. The problem with this approach is that abnormal behavior can be treated as normal behavior. To overcome the limitations of the existing methods, machine learning algorithms have been integrated to improve the efficiency of attack detection. The machine learning algorithms are broadly classified as supervised and unsupervised learning. The former approach, given a set of historical data along with class labels, constructs a model. The model can later be used to classify new attacks. The latter approach is suitable when the data set is not having predefined labels. It clusters the data sets. When a new attack arises, it finds its corresponding cluster.

The objective of this chapter is to provide a brief description about cybersecurity, various cyber attacks, goals of the security system, traditional threat detection algorithms and its drawbacks, Machine learning algorithms for threat classification and prediction.

ATTACK AND ITS CLASSIFICATION

An Attack/ intrusion is defined as any set of actions that tends to compromise cryptographic pillars such as integrity, confidentiality, authentication (Khan, L., et al., 2007, pp. 507). An attack bypasses the security mechanism of a network ((Kaur, R., et al., 2016), (Iwendi, C., et al., 2018, pp. 47258)). Such attacks are classified into the Active attack and Passive attack.

Active Attack

Active attack is the one where the intruder changes the data present in the system. Some examples of active attacks are Denial of Service attack, Man in the Middle attack and Masquerade attack are discussed briefly in this section.

Denial of Service

Denial of service (DoS) is a cyber attack (Gu, Q., and Liu, P. 2007, pp. 454), where the attacker makes the system's resources unavailable for legitimate users. DoS attack consumes resources of various network elements such as routers, firewalls, servers, bandwidth, CPU cycles and thus affects their proper functioning and optimal provisioning of services ((Long, N., and Thomas, R., 2001, pp. 648), (Kurt, B., et al., 2018, pp. 48)). It is of two types: i) Direct attack and ii) Reflector attacks (Chang, R. K., 2002, pp. 42). Direct attacks are the one in which attacker directly sends a large number of packets to the victim system thereby consume resources of victim system. The direct attack includes i) TCP SYN Flooding ii) ICMP SMURF Flooding iii) UDP Flooding. TCP SYN Flooding is otherwise termed to be SYN Flooding. It works as: intruder or hacker keeps on sending TCP connection request with a spoofed source address to the target machine. The target machine instantiates data structures for each request. Once the resources in target machine are exhausted, the device starts to reject further TCP connections, even though the demands are from legitimate users, since the computer has no more resources to satisfy the user request. ICMP SMURF Flooding works as ICMP is used to know whether the system connected to the internet is responding or not. This is achieved by sending echo request. If the system is active, it will send echo reply. Thus attacker broadcast echo request with source address as victim's address and destination address as the broadcast address of the network. Hence all the systems in the network send echo reply to victim system, thereby consuming link bandwidth and degrading performance. UDP Flooding works as: Large UDP packets are forwarded to victim's system, thus congesting the victim's network (Schuba CL et al. 1997, pp. 208). Reflector attacks are otherwise known as indirect attacks. This type of attack uses intermediate nodes like a router, firewall, the gateway for attacking victim (Chang, R. K., 2002, pp. 42). One example of DoS attack is Packet dropping attack. Packet dropping attack is a most threatening denial of service attack where a network element rather than forwarding the packet, starts to discard, thus minimizing the performance of network (Rmyati, M., 2017, pp. 53).

Man-in-the-Middle Attack

An intruder lies between the source and destination host and listens to the communication silently. Sometimes the intruder may either impersonate as source/destination to send or receive messages intended for legitimate communication (Goyal, P., et al., 2010, pp.11). Man in the middle attack cause a severe security threat, and it is more challenging since the attack can be made from a remote location via fake address. It makes use of weakness in authentication protocol used by source and destination since third parties usually provide authentication. The attack can be caused due to i) ARP cache poisoning ii) DNS Spoofing iii) Session Hijacking iv) SSL Hijacking. ARP Cache Poisoning where the attacker invades the network and controls network elements such as a router to monitor network traffic and to get ARP packets sent between source and destination there by starts MIM attacks. DNS Spoofing is an online Man in the Middle attack, where the attacker receives the target's credentials by creating a fake website. When a host system wants to access a web service, it sends DNS request to DNS server, but attacker in between the host system and web service sends the fake address of web service as DNS reply. The host system thinks that the response is from original web service and thereby starts to communicate to counterfeit address and lose all information. Session Hijacking is one, where the data exchanged is obtained using cookies that was to used to establish a session. SSL Hijacking is where the attacker uses fake SSL certificate to impersonate (Gangan, S., 2015).

Masquerade Attack

Masquerade attack is an active attack in which user of a system illegitimately have or assumes the identity of another legitimate user. A masquerader can get access to authorized user's account by stealing credentials or using keylogger. In both cases, the user's identity is acquired illegally (Salem, M. B., and Stolfo, S. J., 2009). A masquerader is an attacker, after successfully getting the users credentials, attempts to perform a malicious action (Tapiador, J., 2011, pp. 297). Masquerade attack can be detected using User behavior profiling and Decoy technology. The technique behind user behavior profiling is that a legitimate user knows the location of the file in a system, which means they are known to the system. Thus user search behaviors are profiled. Whenever an intruder makes use of the system, the abnormal search behavior will be identified as masquerade attack. In Decoy technology, a trap is placed in a file system. The traps are decoy files. A legitimate user puts the decoy files secretly in a system. A malicious user, being unaware of the system and tries to access the decoy files, then masquerader will be trapped (Stolfo, S. J., et al., 2012, pp. 125).

Passive Attack

The passive attack does not disrupt the network, but it snoops the information exchanged over the web, without modifying it, thereby violating the requirement of confidentiality. Therefore, it is quite tricky to detect passive attack (Jawandhiya, P. M., 2010, pp. 4063). Passive attack monitors the network, whereas active attack aims to modify the data exchanged in communication between parties. Examples of passive attacks such as Traffic analysis and eaves dropping are described here.

Traffic Analysis

Traffic analysis is a passive attack which monitors the packet transmission to capture source, destination address (Jawandhiya, P. M., 2010, pp. 4063). It can also be used to determine location and identity of communicating parties.

Eavesdropping

Eavesdropping is a passive attack which aims to obtain some confidential information exchanged between source and destination (Jawandhiya, P. M., 2010, pp. 4063). An eavesdropper intercepts a message transmitted between two communicating parties. Encryption of data before sending is the only way to protect it.

REQUIREMENTS OF A GOOD SECURITY SYSTEM

Security is a prime concern in day to day life. Computer network's security becomes a critical issue, and it is essential to develop mechanisms to defend against the intrusions. A good security system must satisfy the requirements such as authentication, integrity, availability, Nonrepudiability, and confidentiality.

Authentication means assurance that the information is from the source ((Jawandhiya, P. M., 2010, pp. 4063), (Siddiqui, I. F., et al., 2018, pp. 1)). In other words, it is the mechanism used to confirm that a communication request is from an authorized user. There are two types of authentication i) Entity authentication and ii) Data origin authentication. Entity Authentication is justifying the identities of the parties involved in the communication. Data origin authentication focuses on confirming the identity of the creator of the message (Shiu, Y. S., et al., 2011). Integrity is giving assurance that the information transmitted across the network is not altered (Jawandhiya, P. M., 2010, pp. 4063). The term availability represents that services should be available all time, whenever there is a need. The service should be available despite Denial of service attack (Jawandhiya, P. M., 2010, pp. 4063).

Nonrepudiation ensures that source or destination should never deny about sending or receiving the message (Jawandhiya, P. M., 2010, pp. 4063). A digital signature, which serves as a unique identifier for individual users, is used for nonrepudiation purpose (Shiu, Y. S., et al., 2011). The term confidentiality refers to protection of one's own information from unauthorized entities (Jawandhiya, P. M., 2010, pp. 4063). It is closely related to data privacy, such as encryption and key management. Encryption is the standard technique to ensure confidentiality (Shiu, Y. S., et al., 2011).

Intrusion Detection System

The increase in the number of cyber attacks requires the need for security mechanism. One such security mechanism is Intrusion Detection System ((Denning, D. E., 1987, pp. 222), (Koc, L., et al., 2012, pp. 13492), (Arul, R., et al., 2017, pp. 2645)). Intrusion Detection System (IDS) aims to detect various cyber attacks by viewing data records of the same network (Khan, L., et al., 2007, pp. 507). An efficient IDS examines incoming and outgoing activities on the web and thereby identifying malicious activity known as cyber attack, where an intruder tries to compromise the security of the system ((Liu, Z., 2011, pp. 256), (Kabir, E., et al., 2018, pp. 303)). Intrusion detection is a vital tool for the protection of Information and Communication technologies. The term Intrusion detection is a method of detecting actions that an attacker performs against ICT. Such malicious activities are known as an attack, which aims to achieve unauthorized access to Information and Communication system. The intruders are of two types viz. internal and external. Internal intruders are legitimate users inside the network with some access privileges but attempt to get nonprivileged access. External intruders are residing outside the web and try to get unauthorized access to the system (Zarpelao, B. B., et al., 2017, pp. 25). The challenges in designing Intrusion Detection System are that the number of false positives alarm rate will increase and ability to recognize an attack is uncertain. A good intrusion detection system should have high precision, recall, as well as low false positive rate and low false negative rate (Mohammed, M. N., and Sulaiman, N., 2012, pp. 313). The variants of Intrusion detection system is Host-based Intrusion Detection system and Network-based Intrusion Detection System. In host-based Intrusion detection system, the data gathered on the system where IDS is installed is alone monitored. Whereas in Network-based Intrusion Detection System, the network traffic data is used to detect intruders (Khan, L., et al., 2007, pp. 507). Network-based Intrusion Detection System is broadly classified into Anomaly Detection system and Misuse Detection system (Santoro, D., et al., 2017, pp. 79) are described in this section.

Misuse Detection System (MDS)

Misuse Detection System is used to find attack from known attack (Khan, L., et al., 2007, pp. 507). The system detects the attack by comparing network traffic with the signature of attack. It is otherwise termed to be signature-based Detection system. Signature-based Intrusion detection system is accurate and efficient. The drawback of Misuse Detection system is that it can't be able to identify attacks which do not belong to set of signature of attacks. If the attack is launched for the first time, then this method can't be able to detect whether the action is attack or normal behavior (Santoro, D., et al., 2017, pp. 79). Misuse detection system uses learning algorithm which is trained by dataset to classify cyber attack and normal attack. The algorithm does not detect the new novel attack. The algorithm can be retrained with new instances as the training set (Koc, L., et al., 2012, pp. 13492). Z-Wave Networks are networks implemented based on International Telecommunications Union-Telecommunication Standardization Sector (ITU-T) G.9959 recommendation, which specifies a short range, narrow band, sub-gigahertz communication, which is a fast-growing protocol in existence at the moment. It is challenging to evaluate the security of Z-Wave Networks due to the NDA that the developers working on them are required to sign. There already exists an intrusion detection system based on Z-Wave Networks – Misuse-Based Intrusion Detection System (MBIDS) but it only has a 92% mean detection rate, and it cannot detect valid packets injected by an attacker attempting to disrupt normal network operation (Fuller, J. D., et al., 2017, pp. 44).

Anomaly Detection System (ADS)

Anomaly Detection System is a method of searching malicious behavior from normal behavior pattern (Khan, L., et al., 2007, pp. 507). The system compares the network traffic with normal behavior. Anomaly detection system builds a model of normal network behavior and detects any action as malicious if it does not match with the model (Koc, L., et al., 2012, pp. 13492). It is not accurate as Misuse based detection system since it does not perform well for the false positive ratio. But the prime advantage of Anomaly Detection System is its capability to identify the malicious action as an attack even if it occurs for the first time ((Santoro, D., et al., 2017, pp. 79), (Koc, L., et al., 2012, pp. 13492)). Anomaly detection in wireless sensor networks is an important challenge for tasks such as fault diagnosis, intrusion detection, and monitoring applications. Anomaly Detection is the set of data points that are considerably different than the remainder of data. This article analyzes state of the art in anomaly detection techniques for wireless sensor networks and discusses some open issues for research. Several applications of anomaly detection in the context of sensor networks have been described. The strengths and weaknesses of the general approaches are compared. Anomaly detection is used when it is important to detect abnormal behavior without knowing a priori what the abnormality should look like. The main idea is to detect the abnormality even before knowing how the abnormality will be and this is where anomaly detection is used. In general, anomaly or outlier detection mechanisms can be categorized into three general approaches depending on the type of background knowledge of the data available. In this article, the main focus is on two techniques The Statistical Technique and The Non-Parametric Technique. In statistical techniques, the density distribution of the data points being analyzed for anomalies is known theoretically. Non-parametric anomaly detection techniques do not assume any prior knowledge about the distribution of the data. This article also discusses various non-parametric approaches that are used in sensor networks they are rule-based, CUSUM, data clustering, density-based, and support vector machine approaches. Finally, all the approaches are compared. The comparison of approaches to anomaly detection and open issues are discussed. It concludes by saying that, a distributed approach with collaboration among nodes, opens up new issues such as how often the nodes should communicate, and effective ways to achieve a consensus decision regarding the detected anomalies while minimizing energy consumption in the network (Rajasegarar, S., et al., 2008). A combining classifier model based on tree-based algorithms is used for network intrusion detection. The NSL-KDD dataset is used to evaluate the performance of the detection algorithm. The algorithm is used to classify whether the incoming network traffics, normal or an attack. The combining classifier approach is being proposed in this paper to yield better results than the individual classifier. The two classifiers used in this paper are NBTree (Naïve Bayes) and random tree classifiers. The accuracy of the classifiers individually is calculated and compared with the accuracy of the classifiers combined, along with the combining classifier schemes such as the sum rule, product rule, max rule, min rule, median rule, and majority voting have been developed. The result shows that combining classifier approach based on the sum rule scheme can yield better results than individual classifiers, although the opposite is also possible (Kevric, J., et al., 2017, pp. 1051). A general-purpose method called conditional anomaly detection for taking differences among attributes into account is proposed for three different expectation-maximization algorithms for learning the model that is used in conditional anomaly detection. Experiments with more than 13 different data sets compare the algorithms with several other more standard methods for outlier or anomaly detection. The Gaussian Mixture Model, Conditional Anomaly Detection, Detecting Conditional Anomalies, The Direct-CAD Algorithm, The GMM-CAD-Full Algorithm, The GMM-CAD-Split Algorithm, were discussed and compared and the results were discussed. What the results do imply is that, if the goal is to choose the method that is most likely to do better on an arbitrary data set, then the

CAD-GMM-Full algorithm is the best choice. A rigorous, statistical approach is proposed where the various data attributes are partitioned into environmental and indicator attributes, and a new observation is considered anomalous if its indicator attributes cannot be explained in the context of the environmental attributes (Song, X., et al., 2007, pp.631).

INTRUSION PREVENTION SYSTEM

In recent year, with the ever-increasing popularity of the internet, security of networks is a challenging issue for researchers. Intrusion detection system aims to detect cyber attack for normal behavior or pattern, but it is quite difficult to distinguish an attack from normal behavior.

Statistical Algorithm for Threat Detection

The need to build a solution for intrusion in network coined to think of machine learning technologies for preventing intruders. Statistical algorithm relies on distribution of data, prior to constructing a model. Various machine learning methodologies such as Decision trees, Bayesian network, Naïve Bayes, Support vector machine, Linear regression, Logistic regression, Gradient Boosting Regression are usedto protect the network from attacks by analyzing the distribution of data.

Supervised Learning

Supervised learning represents the idea of mapping one or more input variables with the output variable, and the mapping is used to find the output for unseen input variable. The goal of supervised learning is to devise a function $f : x \rightarrow y$, the classifier will learn from the training data and construct a model which is used for testing the unseen data. In general, if the class label is known for an instance then the learning is termed to be supervised learning. Some of the supervised learning methods such as Decision trees, Bayesian Network, Naive Bayes, Support Vector Machine, and Neural Network are briefly described in this section.

Decision Tree

The training data D is represented as a pair $\{x_i, y_i\}$ where each x_i represents the tuple consists of a list of attributes or features along with the values associated with the feature and each y_i represents the class label associated with the tuple. x_i is represented as $\{A_1, A_2, ..., A_k\}$ where each A_i represents the attribute. Each attribute is assigned with more than one values v_i. Depending on the values taken by y_i, the classification problem can be viewed as binary classification and multi-level classification. In the binary classification problem y_i takes only two values (classes), i.e., $y_i \in \{0,1\}$. Whereas in multi-level classification y_i takes more than two values (categories). The decision tree is a binary tree, where each internal node will have zero or two children. The tree is constructed in a top-down fashion, starting from root to leaf and it is finite. Once the tree is built using training set, to test for a tuple, the algorithm does a greedy search through the tree without backtracking. The construction of tree involves two metrics i) Entropy ii) Information Gain. The term Entropy represents the measure of uncertainty associated with an attribute. The value of entropy ranges between [0,1]. Entropy for the data set D is represented in Equation 1 The metric Information gain is used for selecting the attribute. At each level of the tree, always choose the attribute which is having highest Information gain. Information gain for each attribute A_i is represented in Equation 2.

$$Entropy\left(D\right) = \sum_{i=1}^{|c|} -P_i \log_2 P_i \tag{1}$$

$$Info\_Gain\left(D, A_i\right) = Entropy\left(D\right) - \sum_{j=1}^{|v|} P_{v_j} * Entropy\left(D_j\right) \tag{2}$$

Equation 3 and 4 poses P_i, the probability of the ith class and P_{v_j} the probability of the jth value for any attribute A_i respectively.

$$P_i \leftarrow \frac{\#\,tuples\,having\,class\,i}{|D|} \tag{3}$$

$$P_{v_j} \leftarrow \frac{|D_j|}{|D|} \tag{4}$$

The Decision tree to identify the nodes which are intruders is depicted in Figure 1. Decision tree is ideal learning model for designing Intrusion Detection System (Markey, J., & Atlasis, A., 2011). The Decision tree is suitable for handling large data sets. Since the flow of data across networks is enormous in real time, it is worth to use decision tree for such systems. The property that makes Decision tree ideal for Intrusion Detection System is Generalization Accuracy. In the modern era, new attacks are keeps on emerging, but these attacks are small variations of known attacks. Decision trees are efficient in detecting these attacks, due to the property Generalization Accuracy. The dataset used is KDD 1999, prepared by DARPA Intrusion Detection Evaluation Program by MIT Lincoln Labs. The results revealed that Decision tree has high classification accuracy than SVM (Peddabachigari, S., et al., 2004, pp. 118). The Decision tree can be constructed using data present in Honeypots. Decision trees are useful in the analysis of large sets of intrusion detection data. The Decision tree produces a set of rules that are transparent, easy to understand and can be integrated into firewalls (Markey, J., & Atlasis, A., 2011). Decision tree split is used for signature-based intrusion detection. The NSL KDD dataset is used for comparing Decision tree split with C4.5, CART. The split value is calculated using the average of all the values in the domain of an attribute. The efficiency of an algorithm depends on data size and the number of attributes (Rai, K., et al., 2016, pp.2828). Improved decision tree using binary split and the Quad split is used to detect probe, U2R, and R2L attacks. KDD 99 dataset is used to compare Decision tree with binary split and quad split with ID3 (Puthran, S., & Shah, K., 2016, pp. 427).

Bayesian Network

A Bayesian network uses the principle of a joint probability distribution of a set of attributes $A \leftarrow \left\{A_1, A_2 ..., A_{|A|}\right\}$ as a directed acyclic graph and a conditional probability distribution table. Each node corresponds to an attribute, and the conditional probability table associated with it gives the probability of each state of the attribute, given every possible combination of states of its parents ((Koc, L., et al., 2012, pp. 13492), (Lowd, D., & Domingos, P., et al., 2005, pp. 529)). The Bayesian network otherwise termed to be probabilistic inference system tends to compute the posterior probability distribution for a set of query attributes, given some observed values for the attributes. The Figure 2 represents the Bayesian network with five attributes along with conditional probability distribution. Let x_u represents the query attribute. E represents the set of evidence variables, $\left\{E_1, ... E_k\right\}$ and e is a particular observed event. Let Y denotes the hidden variable. The objective is to find the posterior probability distribution $P\left(x_u \mid e\right)$. The posterior probability can be computed by a sum of products of conditional probability from the network. The computation of posterior probability in Bayesian network is $P\left(x_u \mid e\right) = \alpha P\left(x_u, e\right) = \alpha \sum_y P\left(x_u, e, y\right)$

Figure 1. Continuous Decision Tree for Intrusion Detection

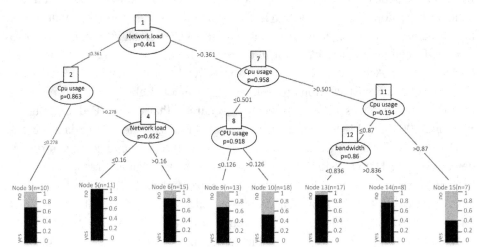

Consider the query, $P\left(A_1 = T \mid A_4 = T, A_5 = T\right)$. The query can be re-written as

$$P\left(A_1 \mid A_4, A_5\right) = \alpha * P\left(A_1, A_4, A_5\right) = \alpha * \sum_{A_2} \sum_{A_3} P\left(A_1, A_2, A_3, A_4, A_5\right) \qquad (5)$$

Equation 6 represents the rewritten of the Equation 5

$$P\left(A_1 \mid A_4, A_5\right) = \alpha \sum_{A_2} \sum_{A_3} \left(P\left(A_1\right) * P\left(A_2\right) * P\left(A_3 \mid A_1, A_2\right) * P\left(A_4 \mid A_3\right) * P\left(A_5 \mid A_3\right)\right) \qquad (6)$$

Equation 7 and Equation 8 represents the computation of α

$$\alpha = \frac{1}{P\left(A_4, A_5\right)} \qquad (7)$$

$$\alpha = \frac{1}{\sum_{A_1, A_2, A_3} \left(P\left(A_1\right) * P\left(A_2\right) * P\left(A_3 \mid A_1, A_2\right)\right) * P\left(A_4 = T \mid A_3\right) * P\left(A_5 = T \mid A_3\right)} \qquad (8)$$

Figure 2. Bayesian Network With 5 Attributes

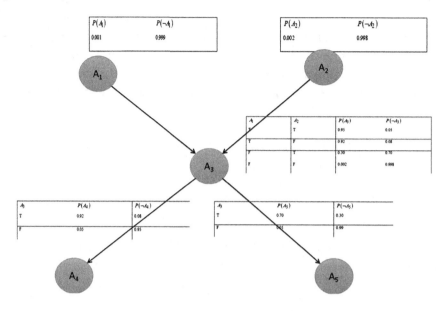

The time complexity of a Bayesian network with $|A|$ attributes is $O\left(2^{|A|}\right)$. The Bayesian filter is used to determine whether the node is malicious or not in Vehicular Adhoc Network. Bayesian network computes the percentage of packets that are not forwarded by the detected node. If this percentage is more significant than the threshold, then it outputs that the identified node is malicious. Otherwise, the node is not malicious (Rupareliya, J., et al., 2016, pp. 649). The Bayesian network does not take into account of the state and time-dependent properties (Gribaudo, M., et. al., 2015, pp. 91). Bayesian approach is used to detect a sudden change in Session Initiation Protocol (SIP) traffic. It is based on hierarchical probability model called hidden Markov model that relates the known, observed features gathered from network traffic to hidden variables. The multiple change-point model is one, where the known variables are dependent on latent states and states either follow previous one or jump to a new state. Bayesian multiple change-point model is used to detect DDoS flooding attacks in VoIP networks (Kurt, B., et al., 2018, pp. 48). Bayesian inference is used to identify the malicious node in Mobile Smartphone Networks (MSN). Bayesian Inference is a statistical method which uses Bayes theorem to update the probability estimate for a hypothesis, used to find the trust value for MSN node (Meng, W., et al., 2017, pp. 162).

Naive Bayes

The naïve Bayes classifier assumed that all the attributes/features/events are independent given a class label. It uses Bayes theorem of probability to identify the class label for the unknown data set. This learning model is easy to build, and it is suitable for handling large data sets. The Equation 9 represents Bayes theorem used to predict class label.

$$P\left(c_i \mid \left(v_j\right)_{A_i}\right) = \frac{P\left(\left(v_j\right)_{A_i} \mid c_i\right) * P\left(c_i\right)}{P\left(\left(v_j\right)_{A_i}\right)} \tag{9}$$

The Table 1 represents the calculation of probability represented in Equation 9. Given a tuple x_u, the naïve Bayes classifier will predict that x_u belongs to the class c_k, if the class c_k is having highest posterior probability conditioned on x_u. $x_u \in c_k$ if the Equation 10 holds otherwise $x_u \notin c_k$

$$P\left(c_k \mid x_u\right) > P\left(c_m \mid x_u\right) \text{ where } k \neq m \And \And 1 \leq k, m \leq |c| \tag{10}$$

Equation 11 represents the prediction of the class label for the tuple x_u, which involves finding the maximum posterior hypothesis for all classes conditioned on the tuple x_u.

$$P\left(c_i \mid x_u\right) = \frac{P\left(x_u \mid c_i\right) * P\left(c_i\right)}{P\left(x_u\right)} \tag{11}$$

$P\left(x_u\right)$ is a constant for all classes irrespective of any problem. Thus one should maximize the numerator $P\left(x_u \mid c_i\right) * P\left(c_i\right)$. Naïve Bayes strictly follows conditional class independence $P\left(x_u \mid c_i\right)$ is computed in the Equation 12

$$P\left(x_u \mid c_i\right) = \prod_{j=1}^{|A|} P\left(Aj \mid c_i\right) = P\left(A_1 \mid c_i\right) * P\left(A_2 \mid c_i\right) * \dots * P\left(A_{|A|} \mid c_i\right) \tag{12}$$

Finally, the class for a tuple x_u is given in Equation 13

$$class\left(x_{u}\right) = \arg\max\left(\prod_{j=1}^{|A|} P\left(Aj \mid c_{i}\right)\right) 1 \leq i \leq |c| \tag{13}$$

Hidden Naïve Bayes (HNB) model is used for intrusion detection problem when the data set is having huge dimensions, where the dimensions are highly correlated, and the network traffic volume streams continuously. The HNB model is having high accuracy, error rate, and also misclassification cost is high. Further HNB is having excellent predictor accuracy than SVM. The intrusion detection model is multinominal classifier which classifies network events as normal or attacks such as DoS, U2R, R2L, Probe. The dataset used is KDD 99, which contains continuous values. Since HNB is a discrete classifier, the continuous values are converted into discrete values using entropy minimization discretization and proportional K-interval discretization. The HNB model has excellent performance in detecting Denial of Service attack (Koc, L., et al., 2012, pp. 13492). In Naïve Bayes classifier, the probability of one attribute does not affect the probability of other. Feature vitality based reduction method is used to reduce the dimensions of NSL KDD data set. Naïve Bayes is used as a classifier that validates the Feature vitality based reduction method (Kaur, R., et al., 2016).

Support Vector Machine

Support Vector Machine is a useful method for classifying the data. The goal of SVM is to create a learning model based on training data which predicts the class label

Table 1. Computation of Probability

Attribute \ Class	c_1	...	c_k	$P\left(v_j \mid c_i\right)$	$P\left(v_j\right)$
v_1	$n\left(v_1 \mid c_1\right)$	...	$n\left(v_1 \mid c_k\right)$		
v_2	$n\left(v_2 \mid c_1\right)$	...	$n\left(v_2 \mid c_k\right)$	$\dfrac{n\left(v_j \mid c_i\right)}{n\left(c_i\right)}$	$\dfrac{\sum_{i=1}^{k} n\left(v_j \mid c_i\right)}{\sum_{i=1}^{k} n\left(c_i\right)}$
...	...	...	...		
v_n	$n\left(v_n \mid c_1\right)$	...	$n\left(v_n \mid c_k\right)$		
$n\left(c_i\right)$	$\sum_{j=1}^{n} n\left(v_j \mid c_i\right)$				

for the test data (Hsu, C. W., et al., 2003). SVM is suitable for linearly separable binary sets. The idea behind the SVM is that input vectors are non-linearly mapped to a very high dimensional feature space (Cortes, C., & Vapnik, V., 1995, pp. 273). The goal of SVM is to design an optimal hyper plane that classifies all the training vectors into two classes. There can be more than one hyper plane that separates the training data into two sets. But to design, a hyper plane that leaves the maximum margin for both classes is challenging task. The margin is the distance between the hyperplane and the closest elements from this hyper plane. Equation 15 represents the hyper plane. Figure 3 represents the hyper plane separating two classes.

$$g\left(\vec{x}\right) = \vec{w}^T\vec{x} + w_0 \tag{15}$$

$$g\left(\vec{x}\right) \leftarrow \begin{cases} \geq -1 & \forall \vec{x} \in Class1 \\ \leq 1 & otherwise \end{cases} \tag{16}$$

Equation 17 represents the computation of the distance between any point and the hyper plane

$$z = \frac{\left|g\left(\vec{x}\right)\right|}{\left\|\vec{w}\right\|} = \frac{1}{\left\|\vec{w}\right\|} \tag{17}$$

Equation 18 represents the computation of total margin from either side of the hyper plane.

$$\frac{1}{\left\|\vec{w}\right\|} + \frac{1}{\left\|\vec{w}\right\|} = \frac{2}{\left\|\vec{w}\right\|} \tag{18}$$

Minimizing the term $\frac{2}{\left\|\vec{w}\right\|}$ will maximize the separability. The minimization of $\vec{w}$ is a nonlinear optimization and thus solved using Kuhn-Tucker conditions using Lagrangian multiplier and is represented in Equation 19

$$\vec{w} = \sum_{i=1}^{|D|} \lambda_i y_i \vec{x_i} \tag{19}$$

A novel statistical technique for intrusion detection system is performed in two stages using Least Square support vector machine (LS-SVM). In the first stage of decision making, the whole data set is divided into some predetermined arbitrary subgroups. The algorithm helps in selecting representative samples from entire dataset and optimum allocation scheme is developed based on the observation within these subgroups. In the second stage of decision making, the LS-SVM is applied to detect intrusions. The experiments are carried out on KDD 99 data set. All binary- classes and multi-classes are tested that obtains accuracy and efficiency (Shah, S. A. R., & Issac, B., 2018, pp. 157). Suricata Intrusion Detection system could process better at high speed of network traffic with lower packet drops, but it consumes more resource. Snort Intrusion detection system had higher detection accuracy but with high rate of false positive alarms. The hybrid version of support vector machine and Fuzzy logic produce better detection. IDS designed using SVM helps to detect malicious traffic, undetected malicious traffic, legitimate traffic that uses IDS identify as malicious and legitimate traffic that IDS detects as good. Commercially available IDS are McAfee, Cisco, Symantex etc. Some open source IDS are snort, Suricata and Bro. Snort and Suricata support Intrusion prevention system. Suricata is more scalable with multi-threading architecture and IPV6 protocol support. Snort and Suricata are efficient with network traffic of up to 10Gbps. With increase in network traffic, there is possibility of packet loss that leads to induce of malicious over network. Snort and Suricata uses pre-defined rules to identify malicious activities. So any unmatched pattern will not be detected by it. The reason behind is snort, and Suricata does not use any machine learning algorithm (Kabir, E., et al., 2018, pp. 303). Clustering Tree based on SVM (CT-SVM) is proposed in order to minimize running time of SVM. The CT-SVM is applied on network-based anomaly detection. The dataset used is 1998 DARPA. Network traffic had been captured using packet capturing utilities such as libpcap or operating system call and it is stored as data in audit file. Each instance can be either an attack or normal behavior. Four categories of attack are described as Denial of service, U2R, R2L, and probe. Number of classes and features are 5 and 41 respectively. Some attributes take continuous values and some take nominal values. Nominal values should be converted into continuous values for classification and clustering algorithm. LIBSVM is used to implement SVM. The proposed CT-SVM produces better accuracy, false positive rate and false negative rate. Clustering is used to reduce training set and to optimize support vectors (Khan, L., et al., 2007, pp. 507). SVM is suitable for data sets having two classes. It is insensitive when the instances increase. In a data set, it is quite reasonable that some feature may have relatively large value when compared to other features; Normalization is used to handle differences in values among features. SVM does not require a reduction in the number of features, to avoid over fitting. SVM also has the advantage of low probability generation of error. These benefits make SVM to be used in intrusion detection. The data used for training SVM are KDD 99. The experimentation is carried using LIBSVM Package. Normalization is integrated into SVM. Without performing normalization, SVM spends most of its time in calculation (Liu, Z., 2011, pp. 256). Among the other learning methodologies, SVM is best tool for small sample learning and is efficient for network attack classification. The data set used for experimentation is KDD cup 99. Each instance is having 41 features. Gradual feature reduction is used to reduce number of features from 41 to 19. The redundancy in KDD99 data set is high. Thus K-Means clustering is used to select instance with no duplication and thereby reducing the size of data. SVM is used as a classifier which yields 98.6249% of accuracy(Li, Y., et al., 2012, pp. 424). SVM depends on characteristics of training data but not on dimensions of features (Mohammed, M. N., and Sulaiman, N., 2012, pp. 313).

Neural Networks

Neural networks are statistical learning methods and are used to solve the non-linear problem. A neural network is a system which consists of inputs and outputs and is integrated with simple and similar processing elements. Each processing element

Figure 3. Construction of hyper plane

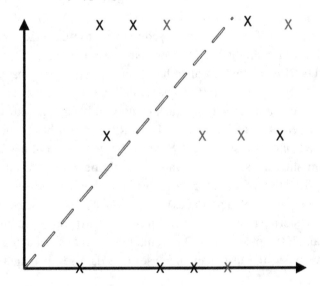

is associated with weight. If the weight associated with the processing element changes, then it changes the behavior of the whole network. The objective is to choose the appropriate weight to achieve prompt input-output relationship. This process is termed to be training the network. If the weight is kept constant, the output depends only on current input and independent of past inputs. Thus the network is termed to be memory less (Nguyen, D. H., & Widrow, B., 1990, pp. 18). Figure 4 shows the representation of a neural network with 2 input units, 3 hidden units, and one output unit. It involves forward propagation and backward propagation. Forward propagation consists in assigning random weights to each connection from the input unit to hidden unit and from hidden unit to output unit. Backward propagation involves measuring error, which is the difference between the target output and the calculated output and accordingly adjusts the weights to decrease the error. Both forward and backward propagation is repeated till the output is predicted accordingly. Let the initial value of the ith unit in the input layer is v$_i$. The weight from the ith input unit to kth hidden unit is given as w$_{ik}$. Equation 20 and Equation 21 represents the calculation of values in each hidden layer unit. The intermediate values in the hidden layer can be calculated as the sum of the product of the values from each input unit and its corresponding weight. The activation function is applied to the intermediate value, which should be given as input to the output unit. The activation function can be linear, sigmoid or hyperbolic tangent. Equation 22 and Equation 23 represents the calculation of intermediate value and the final value of the output unit. Equation 24 represents the calculation of error in the output unit. In order to minimize the error rate, the deviation in sum is calculated in Equation 25.

$$H_{sum_k} = \sum_{i=1}^{n} v_i * w_{ik} \tag{20}$$

$$H_{Res_k} = g\left(H_{sum_k}\right) \tag{21}$$

$$O_{sum_j} = \sum_{i=1}^{n} H_{Res_k} * w_{kj} \tag{22}$$

$$O_{Res_j} = g\left(O_{sum_j}\right) \tag{23}$$

$$Err_j = O_{Res_j} - Actual\ output \tag{24}$$

$$\Delta_{sum_k} = g'\left(O_{sum_j}\right) * Err_j \tag{25}$$

Now, differentiate Equation 22 with respect to weight from hidden unit to the output unit.

$$\frac{dO_{sum_k}}{dw_{kj}} = H_{Res_k} \tag{26}$$

$$dw_{kj} = \frac{dO_{sum_j}}{H_{Res_k}} \tag{27}$$

Now, differentiate Equation 22 with respect to the result of the hidden unit.

$$\frac{dH_{Res_k}}{dO_{sum_j}} = \frac{1}{w_{kj}} \tag{28}$$

$$dH_{Res_k} = \frac{dO_{sum_j}}{w_{kj}} \tag{29}$$

Similarly, differentiating the activation function of the hidden layer result,

$$g'\left(H_{sum_k}\right) = \frac{dH_{sum_k}}{dH_{Res_k}} \tag{30}$$

Multiplying the Right hand side of Equation 30 on both sides of Equation 29

$$dH_{Res_k} * \frac{dH_{sum_k}}{dH_{Res_k}} = \frac{dO_{sum_j}}{w_{kj}} * \frac{dH_{sum_k}}{dH_{Res_k}} \tag{31}$$

On simplifying,

$$dH_{sum_k} = \frac{dO_{sum_j}}{w_{kj}} * g'\left(H_{sum_k}\right) \qquad (32)$$

Repeatedly performing forward and backward propagation minimizes the error and achieves maximum accuracy.

Regression

The goal of regression is to construct a mathematical model that explains the relationship between the variable. Let us consider two variables $\left(x_i, y_i\right)$ where x_i is termed to be explanatory variable or regressor or independent variable and y_i is termed to be response or predictor or dependent variable. Regression aims to find a model that best fits all x_i and y_i (Seber, G. A., & Lee, A. J., 2012). For example, consider Newton's second law $F = ma$ where F is a predictor and a is regressor. Section 5.1.2 describes Linear Regression, Logistic Regression, and Gradient Boosting Regression

Linear Regression

Let the input vector is represented as $\left\{x_i \in \mathbb{R}^m\right\}_{i=1..num\_instances}^{m=1..num\_features}$ and the predictor is given as $y \in \mathbb{R}$. The objective is to construct a linear regression model (Nguyen, D., et al., 2011, pp. 115) described in Equation 36. If each instance has one regressor

Figure 4. Representation of Neural Network

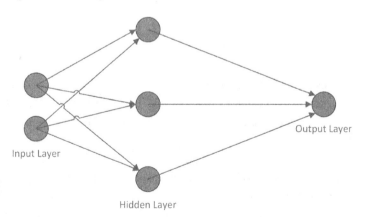

261

and predictor, then it is termed to be simple Linear Regression. If more than one regressor is present, then it is termed to be multiple linear regression.

$$\hat{y} = \alpha + x\beta \tag{36}$$

$\hat{y}$ is the fitted value for the output variable. A straight line is drawn between regressor and predictor, and the error is measured represented in Equation 37 Let the predicted value for instance x_i is, $\hat{y}_i$ and the observed value is y_i. Then the Mean Squared error, for instance, x_i is computed as in Equation 37. The goal of linear regression is to minimize mean squared error represented in Equation 37.

$$e_{x_i} \leftarrow \left(y_i - \hat{y}_i\right) \tag{37}$$

$$MSE \leftarrow \frac{1}{|D|} \sum_{i=1}^{|D|} \left(e_{x_i}\right)^2 \tag{38}$$

Squaring of error in Equation 38 is needed because error can be positive or negative. The intercept α and β are computed as represented in Equation 39 and Equation 40 respectively.

$$\alpha = \bar{y} - \beta\bar{x} \tag{39}$$

$$\beta = \frac{\sum_{i=1}^{|D|} \left(x_i - \bar{x}\right)\left(y_i - \bar{y}\right)}{\sum_{i=1}^{|D|} \left(x_i - \bar{x}\right)^2} \tag{40}$$

In some cases, it is difficult to minimize sum squared error (Ordinary Least Square (OLS)), because of dichotomous nature of output variable for the predictor variables are categorized, and the mean of outcome variables are computed for each category. Thus, if the resultant points are plotted sigmoidal curve will be appeared. In the sigmoidal 'S' shaped curve, it is difficult to apply linear regression, since it is difficult to handle extreme points and errors are neither normally distributed and nor it is constant (Peng, C., et al., 2002, pp. 3). Thus to overcome the drawbacks, a variation of Linear regression called Logistic regression is proposed. Multivariate regression is used for detecting the intrusions. The experiment is carried over KDD

Cup 1999 data and the data is divided into a set of training vectors and test vectors since the original data is too large. The results show that multivariate regression achieves high accuracy with a very high speed of detection (Bernick, J. P., 2007). Multivariate linear regression is used for designing anomaly detection system in order to detect abnormal behavior from normal behavior. The experiment is carried over the log file gathered in a personal computer. The attributes taken into account are Demand zero faults/s, Cache Bytes, page faults/s, Transition Faults/s, Pool pages allocs, Committed bytes in use, Available MByte, Cache Faults/s, Write copies/s, Pages/s. The experimental result shows that the multivariate regression detects the outlier efficiently (Gautam, S. K., & Om, H., 2015, pp. 361).

Logistic Regression

Logistic regression is a statistical method that predicts the independent variable y from one or more dependent variable x_i. If the independent variable y takes only two values, i.e., binary or dichotomous, then the logistic regression is termed to be binomial or binary logistic regression. If the independent variable y has more than two values, then it is termed to be multinominal logistic regression. Logistic regression works well for categorical outcome variable (independent or response variable) and one or more categorical or continuous (dependent or explanatory variable of predictor variables). Logistic regression solves the problem of linear regression by applying natural logarithm on dependent variable (Peng, C., et al., 2002, pp. 3). Two prime concept in logistic regression is i) odds and ii) probability. Probability is the ratio of the number of favorable cases to the total number of events in sample space. Odds are the ratio of probability, i.e., the ratio of the probability of an event for the favorable outcome to the probability of the event not favorable for the outcome (Sperandei, S., 2014, pp. 12). Let the dependent variables are represented as $\{x_1, x_2, \ldots x_m\}$ and the independent variable y is dichotomous, i.e., $y \in \{0,1\}$ and logistic of y is computed in Equation 41.

$$logistic\left(y\right) = \ln\left(\frac{p}{1-p}\right) = \alpha + \sum_{i=1}^{m} \beta_i x_i \tag{41}$$

where,

α is termed to be intercept parameter

β_i is termed to be regression coefficient associated with dependent variable estimated by Minimum likelihood method.

p indicates the probability of an event which can be computed as in Equation 42

$$p = P\left(Y = 1 \mid x_1, x_2, \ldots x_m\right) = \frac{e^{\alpha + \sum_{i=1}^{m} \beta_i x_i}}{1 + e^{\alpha + \sum_{i=1}^{m} \beta_i x_i}} \tag{42}$$

Multinominal logistic regression is used to detect anomalies from usual behavior. The KDD-cup 1999 data is used for experimentation and the attack types such as probe, DoS, U2R and R2L are considered as independent variable for logistic regression. The logistic regression achieves high sensitivity and specificity as well as low false positive and false negative rates (Wang, Y., 2005, pp. 672). Scan detection using Bayesian logistic regression analyze all the traffic during a specified time interval originating from each source. The purpose of logistic regression is to determine the probability of the traffic from a source is having scanning activity. Scanning activity now a days tends to gather user's information. Bayesian logistic regression is compared with Threshold random walk, the former achieves high accuracy than latter (Gates, C., et al., 2006, pp. 402).

Gradient Boosting Regression

Gradient Boosting Regression is a statistical method that aims to find the approximate function $\hat{F}\left(x_i\right)$ that estimates the original function $F\left(x_i\right) = y_i$, thereby minimizing the loss function. For regression, the loss function is measured as either squared error represented in Equation 43 or absolute error represented in Equation 44. The term boosting is used to boost up a weak learning model, so that it can function better. Equation 45 represents the loss function.

$$e_{x_i} \leftarrow \left(F\left(x_i\right) - \hat{F}\left(x_i\right)\right)^2 \tag{43}$$

$$e_{x_i} \leftarrow \left|F\left(x_i\right) - \hat{F}\left(x_i\right)\right| \tag{44}$$

$$L\left(F\left(x_i\right), \hat{F}\left(x_i\right)\right) = \left(y_i, \hat{y}_i\right)^2 \tag{45}$$

The negative gradient, which is termed to be residue is computed by taking the derivative of the loss function with respect the approximate function $\hat{F}\left(x_i\right)$. Algorithm 1 describes the working of Gradient Boosting Regression.

$$Residue_{x_i} \leftarrow \frac{\partial L\left(F\left(x_i\right), \hat{F}\left(x_i\right)\right)}{\partial \hat{F}\left(x_i\right)} \tag{46}$$

Gradient Boosting trees (GBT) are used to design a host based intrusion detection system. The information present in the host machine are used to construct gradient boosting trees to detect DoS attacks. DoS attacks are developed in the system which includes L2 Cache DoS, Backside bus bandwidth (BSB) DoS, Front-side bus bandwidth (FSB) DoS, Memory DoS, Loop DoS. The experimentation is carried with the five attacks as class labels and the Intrusion Detection System is trained using Gradient Boosting Trees model. The proposed GBT model achieves high accuracy in detecting DoS attacks (Tao, R., et al., 2009, pp. 13).

Algorithm 1. $GradientBoostingRegression\left(\ \right)$

```
For each  x_i
```

$$F_0\left(x_i\right) \leftarrow \frac{\displaystyle\sum_{i=1}^{n} y_i}{n}$$

```
End For
For each  x_i
        For  m = 1  to  M
```

$$Re\,sidue_{im} \leftarrow \frac{-\partial L\left(F_m\left(x_i\right), \hat{F}_m\left(x_i\right)\right)}{\partial \hat{F}\left(x_i\right)}$$

$$h_m\left(x_i\right) \leftarrow \hat{F}_m\left(x_i\right) - F_m\left(x_i\right)$$

$$\hat{F}_m\left(x_i\right) \leftarrow \hat{F}_{m-1}\left(x_i\right) + \gamma_m h_m\left(x_i\right)$$

```
        End For
End For
```

Non Parametric Algorithm For Threat Detection

Non Parametric algorithms do not rely on the distribution of data. Some of the non parametric algorithms disscussed in this section are K-Means clustering, Hierarchical clustering, Y-Means clustering, Density-based clustering,

Unsupervised Learning

Unsupervised learning is the one where there is no class or target label for an instance. The model will be constructed based on some statistical relationship among the instances. One such example of unsupervised learning is clustering, where the instances will be clustered based on the metrics such as Euclidean distance, manhattan distance and cosine similarity, etc. This section briefly describes some of the unsupervised learning methodologies such as K-Means, Hierarchical clustering, Y-means, Density- based clustering.

K-Means

Given a set of n data sets in d-dimensional space and an integer k, the objective is to identify k points called centers/centroid to minimize the mean squared distance from each data to its nearest centroid. The algorithm works iteratively and tends to find the locally minimal solution (Kanungo, T., et al., 2002, pp. 881). K-Means algorithm is widely used because of its simplicity, scalability, speed of convergence and adaptability to sparse data. The K-Means algorithm is viewed as a gradient descent procedure, which initially chooses random centroid and iteratively updates this centroid to minimize mean squared distance of each data to its centroid represented in Equation 33. K-Means always converge to a local minimum. The local minimum depends on cluster centroid (Oyelade, O. J., et al., 2010).

$$Min \frac{1}{n} \sum_{i=1}^{n} d^2\left(x_i, c_j\right) \tag{33}$$

The $d\left(x_i, c_j\right)$ represents the Euclidean distance between the data point x_i and cluster centroid c_j. Algorithm 2 describes the working of K-Means.

The Network Intrusion Detection Systems (NIDS) defends a network from both insider and outsider intrusions. NIDS is categorized as Misuse Detection (MD), Anomaly Detection (AD). MD based use signatures or patterns of existing attacks, while AD based checks for deviation from normal profiles of the network traffic and reports the attack. NIDS analysis is done for flow based on CIDDS-001 datasets using K-Nearest Neighbors classification and K-Means Clustering algorithm. CIDDS-001 (Coburg Network Intrusion detection Dataset) is a labeled flow-based data set. It is developed to evaluate Anomaly-based NIDS. This dataset contains unidirectional Netflow data. It consists of data from the Open stack and External Servers. CIDDS – 001 data set is selected for Anamoly based NIDS. Simulation is performed using Weka (Waikato Environment for Knowledge Analysis). KNN classifier is used for multiclass classification on Weka. It classifies the instances as normal, attacker, victim, suspicious and unknown. K-Means is used to cluster the instances. Weka is a collection of ML algorithms used for performing data mining tasks. It has various tools for dataset pre-processing, classification, regression, clustering, association and visualization. KNN classifier is built on distance function that measures the similarity or difference between two instances. K-Means algorithm partitions instances into k-clusters in which each instance is associated to the cluster with the nearest mean. K-means and K-NN have the best performance for the metrics TP rate, Detection rate and false alarm rate (Verma, A., & Ranga, V., 2018, pp. 709).

Hierarchical Clustering

Hierarchical, agglomerative clustering is a prominent technique in unsupervised learning. Agglomerative clustering, initially partition the data set into singleton nodes, and in each step, the algorithm tries to merge the current set of mutually closest nodes into a new node, until there is only one node left which comprises

Algorithm 2. $K - Means\left(c_1, c_2, ...c_k\right)$

```
Initialize  c_i ← { }
For each  d_i ∈ D
{
        For each  c_j ∈ c
        {
```
$$dist_{d_i c_j} \leftarrow sqrt\left(\left(x_i - x_j\right)^2 + \left(y_i - y_j\right)^2\right)$$
```
        }
```
$$cid \leftarrow FindMin\left(dist_{d_i c_j}\right)$$
$$c_{cid} \leftarrow c_{cid} \cup d_i$$
```
}
```

the entire data set. The methods to measure Intra cluster dissimilarity are single, complete, average, weighted, ward, centroid and median linkage. The input should be represented as a pairwise dissimilarity between data points, and the output is a stepwise dendrogram (Müllner, D., et al., 2011). The goal of hierarchical clustering is to find a hierarchy of partitioning rather than finding a single partition (Balcan, M. F., et al., 2014, pp. 3831). Figure 5 represents the hierarchical clustering of 14 instances into two clusters.

The hierarchical clustering algorithm is used in conjunction with support vector machine, which aims to perform feature selection from KDD cup 1999 training set. Feature selection using hierarchical clustering reduces the training time and helps to boost the performance of support vector machine (Horng, S. J., 2011, pp. 306).

Y-Means

To overcome the shortcoming of K-means, Y-Means is proposed. The drawbacks of K-Means are i) Dependent on the number of clusters. ii) Degeneracy. K-Means require users to specify the value for 'k,' the number of clusters. But the problem is that users are not aware of the distribution of data. The initial value of 'k' may lead to the poor partitioning of data. Degeneracy represents the scenario when an algorithm returns empty cluster. The Y-means algorithm automatically decides the number of clusters based on the statistical nature of data. The algorithm makes use of initial centroid, and repeatedly assigns instances to cluster centroid. The objective is to find an optimal number of clusters which is independent of the initial centroid.

Figure 5. Representation of Dendrogram

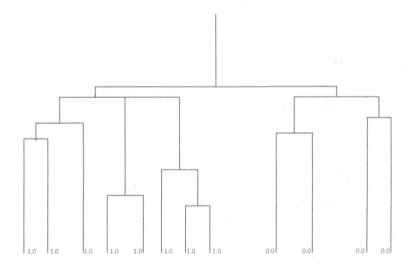

At each iteration, the algorithm does splitting and merging. Splitting is used to remove outliers. At each iteration, for each cluster centroid, a confident area is determined. The Confident area is the circle with radius 2.32 (Guan, Y., 2003, pp. 1083). Any data point lies out of the confident area is termed to be an outlier. The remotest outlier is considered to be new cluster centroid. Some instances from its neighbor will now be assigned to the new cluster centroid. The splitting continuous as long as there is no outlier exist. The instances within the same cluster are more similar to one another, but at the same time number of clusters will increase. Merging is the process of combining two clusters if they overlap each other. Clusters c_i and c_j will be merged if there are some data points that lie in the confident area of c_i and c_j (Corchado, E., et al., 2010, pp. 23). Algorithm 3 describes the working of Y-Means

Y-Means clustering is a combination of K-Means and other clustering mechanisms. It partitions the large data space and also overcomes the drawback of K-means, i.e., one no need to mention the number of clusters. Y-Means is applied on the KDD-99 data set. Y-means achieves a detection rate of 89.89% and false alarm rate of 1.00% (Guan, Y., 2003, pp. 1083).

Density-based Clustering

Density-based clustering algorithm is used for finding the non-linear shaped cluster. A cluster is defined as a maximal number of densely connected points. The algorithm makes use of the concept of density reachability. A data item d_i is said to be densely reachable from d_j, if d_j is in ε the distance from d_i represented in Equation 34 and Equation 35

$$dis \tan ce\left(d_i, d_j\right) \leq \varepsilon \tag{34}$$

$$N\left(d_i\right) \leftarrow \left\{N\left(d_i\right) \cup d_j \mid distance\left(d_i, d_j\right) \leq \varepsilon\right\} \tag{35}$$

A densely connected cluster can be formed if it has a minimum number of data items represented by Min and within ε the neighborhood. The data items are categorized as a core item, an outlier, and border item. A data item d_i represented as $\left(x_i, y_i\right)$ is termed to be a core item if it has more than specified number of Min items within the radius ε. A data item d_i is said to be border item if it has less than a specified number of Min items within the radius. ε An outlier is a data item,

Algorithm 3. $Y - Means(D)$

```
Initialize centroid
```
$c_1, c_2, ...c_k$
$$K - Means(c_1, c_2, ...c_k)$$
```
// check for degeneracy
For each
```
$c_j \in c$
```
If
```
$c_j == \{ \}$
```
then
```

$$\text{Initialize new centroid } c_1, c_2, ...c_k$$
$$K - Means(c_1, c_2, ...c_k)$$

```
        End If
End For
// Check for Outlier
For each
```
$d_i \in D$
```
If
```
$distance(d_i, c_j) > threshold$
```
then
```

$$o \leftarrow o \cup \{d_i\}$$

```
        End If
End For
If
```
$o \neq \{ \}$
```
then
```

$$\text{Initialize new centroid } c_1, c_2, ...c_k$$
$$K - Means(c_1, c_2, ...c_k)$$

```
End If
// Merging
For each
```
$d_i \in D$
```
        For
```
$i \leftarrow 1 \ to \ k - 1$
```
                For
```
$j \leftarrow i + 1 \ to \ k$
```
                        If
```
$d_i \in c_i \ \&\& d_i \in c_j$
```
then
```

$$c_{ij} \leftarrow c_i \cup c_j$$

```
                        End If
        End For
End For
```

which is farthest away from all the points. Algorithm 4 represents working of Density-based clustering.

Density-based Fuzzy imperialist competitive clustering algorithm (D-FICCA) was proposed to handle denial of service attack. Density-based clustering helps imperialist competitive algorithm to control noise as well as forming clusters of arbitrary shape. The D-FICCA handles dataset gathered from sensors implemented at Berkeley Research Lab. Higher detection accuracy of 87% and clustering quality of 0.99 is achieved through DFICCA (Shamshirband, S., et al., 2014, pp. 212). Density and grid-based clustering algorithm are used for detecting anomalies in 1990 KDD cup data set. The proposed Density and grid-based clustering achieves greater benefits regarding computational complexity (Leung, K., & Leckie, C., 2005, pp. 333).

Reinforcement Learning

Reinforcement learning is learning what to do in the environment, how to map situations and actions to maximize reward. The learning involves agents and agents sense the environment in which it is present, thereby taking the actions based on the

Algorithm 4. $DensityBasedClustering\left(\ \right)$

```
Initialize  ε and  Min
Initialize  w_i ← 0 1 ≤ i ≤ |D|
For each  d_i ∈ D
        If  w_i == 0 then
                For each  d_j ∈ D  where  i ≠ j
                    If  distance(d_i, d_j) ≤ ε
                            c_{d_i} ← c_{d_i} ∪ {d_j}
                            w_j ← 1
                    End If
                End For
                If  |c_{d_i}| ≠ Min then
                        o ← o ∪ {d_i}
                End If
        End If
End For
```

interaction in the environment. The important concepts of reinforcement learning are i) policy which maps states to actions ii) reward function numerically evaluate the state, which tells whether the action made by the agent is good or bad. iii) value function which represents the total reward that the agent earns (Sutton, R. S., & Barto, A. G., 2018). Reinforcement learning differs from the supervised learning where the latter requires training data with the class label, and the former does not require so. The former requires that when an agent chooses an action, the reward and its associated state will be informed. But the agent is not notified of which action will take to the best state. The agent must collect all its experiences of state, action, transition and rewards earned to select optimal state. Reinforcement learning problem involves searching for state space tree which has behaviors and finds the one that best fits the environment, which is usually done through the optimization algorithm. Having seen the behavior, the utility value or reward is estimated for the behavior which is done through statistical techniques and dynamic programming (Kaelbling, L. P., et al., 1996, pp. 237). In short, Reinforcement learning is termed to be learning without labels and learning to control the data.

The adaptive neural network which learns new attacks autonomously by using feedback from the system was proposed and designed using reinforcement learning techniques. Autonomous learning of attacks is tested by generating ping flood attack. Cerebellar Model Articulation Controller (CMAC) neural network is a feed-forward neural network designed to produce input, output mappings. CMAC is used for online learning. While training CMAC neural network, the error rate was 97.6758%, which is extremely very high, this is because of random weights. On subsequent training and feedback from the system, the error was reduced to 2.1999% and finally to 1.94^{-7}%. The CMAC neural network is fed with UDP packet storm attack, and the initial response error was 93.2869% but after completing the training of CMAC, the error was reduced to 2.1992%, and it is proved the capability of CMAC neural network to identify a new attack (Cannady, J., 2000, pp. 1).

One example of Reinforcement learning is Q learning, which is discussed in the section 5.3.1

Q Learning

Q learning is a model-free reinforcement learning, where agent's experience is represented by a set of stages. At each stage, the agent chooses an action, which maximizes the reward. The agent in the k state knows following attributes.

- Agent's current state s_k
- The action a_k that agent is going to take in the state s_k

- The learning rate α which can take the value as $0 \leq \alpha \leq 1$. If α is set as 0, then learning does not happen. If it is set closely to 1, then agent learns quickly
- Reward $R_{s_k a_k}$, the agent will get in the state s_k by doing action a_k
- The Discount factor γ which can take the value as $0 \leq \gamma \leq 1$ is used for convergence

Equation 47 represents the computation of Q value for any state.

$$Q\left(s_k, a_k\right) \leftarrow \left(1 - \alpha\right) * Q\left(s_k, a_k\right) + \alpha \left(R_{s_k a_k} + \gamma \cdot \max \left\{Q\left(s_{k+1}, a\right)\right\}_a\right) \tag{47}$$

Game based Fuzzy Q-learning (G-FQL) is used to detect denial-of-service attack in Wireless Sensor Networks. The cost function for G-FQL is the multi objective function which aims to maximize true confidence rate and to minimize false positive and false negative. The experimentation was carried out in NS2 using Low energy Adaptive Clustering Hierarchy (LEACH) protocol. Random function is used to generate the attack. The proposed G-FQL achieves high accuracy in detecting Denial-of-service attacks (Shamshirband, S., et al., 2014, pp. 228). A real time intrusion detection system is designed using Q-learning integrated with rough set theory. The proposed methodology achieves maximum classification accuracy for NSL-KDD data set. The proposed Q-learning has two phases for constructing reward matrix such as initial reward matrix and final reward matrix. Accuracy is discretized in the range [-1, 1, 0] for each action. NSL-KDD data set is used, which consists of 42 attributes. Among 42 attributes, 41 are conditional and 1 is decision. The proposed Q learning methodology achieves maximum classification accuracy than other statistical methods (Sengupta, N., et al., 2013, pp. 161).

CONCLUSION

Cyber security is the predominant research area where the objective is to maximize security by efficiently detecting and preventing the system from intrusion. In this chapter, various kinds of cyber attacks, methodologies to detect the cyber attacks and the machine learning methods to prevent the attacks are elaborated. Two main classes of intrusion detection system such as signature-based and anomaly-based detection are studied in detail. The drawbacks of existing detection approaches are stated and the machine learning approaches to overcome the mentioned drawback and to predict the intrusions are discussed in detail. There are several machine learning mechanisms such as supervised learning, unsupervised learning, regression, and reinforcement learning to solve the problem of intrusion detection and prevention and thereby helps to maximize the security of the system. Finally, the authors want to state that no methodology can completely detect all attacks since there is an intruder who always wants to penetrate through the system. In the meantime, cryptanalyst finds an alternative solution for the new intrusion.

REFERENCES

Arul, R., Raja, G., Bashir, A. K., Chaudry, J., & Ali, A. (2018). A Console GRID Leveraged Authentication and Key Agreement Mechanism for LTE/SAE. *IEEE Transactions on Industrial Informatics*, *14*(6), 2677–2689. doi:10.1109/TII.2018.2817028

Arul, R., Raja, G., Kottursamy, K., Sathiyanarayanan, P., & Venkatraman, S. (2017). User path prediction based key caching and authentication mechanism for broadband wireless networks. *Wireless Personal Communications*, *94*(4), 2645–2664. doi:10.100711277-016-3877-5

Balcan, M. F., Liang, Y., & Gupta, P. (2014). Robust hierarchical clustering. *Journal of Machine Learning Research*, *15*(1), 3831–3871.

Bernick, J. P. (2007). Very Fast Intrusion Detection by Multivariate Linear Regression. *Computer*.

Cannady, J. (2000, October). Next generation intrusion detection: Autonomous reinforcement learning of network attacks. In *Proceedings of the 23rd national information systems security conference* (pp. 1-12). Academic Press.

Chang, R. K. (2002). Defending against flooding-based distributed denial-of-service attacks: A tutorial. *IEEE Communications Magazine*, *40*(10), 42–51. doi:10.1109/MCOM.2002.1039856

E. Corchado, M. G. Romay, & A. M. Savio (Eds.). (2010). Hybrid Artificial Intelligent Systems, Part II*: 5th International Conference, HAIS 2010, San Sebastian, Spain*, June 23-25, 2010, *Proceedings* (Vol. 6076). Springer Science & Business Media.

Cortes, C., & Vapnik, V. (1995). Support-vector networks. *Machine Learning*, *20*(3), 273–297. doi:10.1007/BF00994018

Denning, D. E. (1987). An intrusion-detection model. *IEEE Transactions on Software Engineering*, *SE-13*(2), 222–232. doi:10.1109/TSE.1987.232894

Fuller, J. D., Ramsey, B. W., Rice, M. J., & Pecarina, J. M. (2017). Misuse-based detection of Z-Wave network attacks. *Computers & Security, 64*, 44–58. doi:10.1016/j. cose.2016.10.003

Gangan, S. (2015). *A review of man-in-the-middle attacks.* arXiv preprint arXiv:1504.02115

Gates, C., McNutt, J. J., Kadane, J. B., & Kellner, M. I. (2006, June). Scan detection on very large networks using logistic regression modeling. In *Computers and Communications, 2006. ISCC'06. Proceedings. 11th IEEE Symposium on* (pp. 402-408). IEEE. 10.1109/ISCC.2006.142

Gautam, S. K., & Om, H. (2015). Multivariate linear regression model for host based intrusion detection. In *Computational Intelligence in Data Mining-Volume 3* (pp. 361–371). New Delhi: Springer. doi:10.1007/978-81-322-2202-6_33

Goyal, P., Batra, S., & Singh, A. (2010). A literature review of security attack in mobile ad-hoc networks. *International Journal of Computers and Applications, 9*(12), 11–15. doi:10.5120/1439-1947

Gribaudo, M., Iacono, M., & Marrone, S. (2015). Exploiting Bayesian networks for the analysis of combined attack trees. *Electronic Notes in Theoretical Computer Science, 310*, 91–111. doi:10.1016/j.entcs.2014.12.014

Gu, Q., & Liu, P. (2007). Denial of service attacks. Handbook of Computer Networks: Distributed Networks, Network Planning, Control, Management, and New Trends and Applications, 3, 454-468. doi:10.1002/9781118256107.ch29

Guan, Y., Ghorbani, A. A., & Belacel, N. (2003, May). Y-means: A clustering method for intrusion detection. In *Electrical and Computer Engineering, 2003. IEEE CCECE 2003. Canadian Conference on* (Vol. 2, pp. 1083-1086). IEEE. 10.1109/ CCECE.2003.1226084

Horng, S. J., Su, M. Y., Chen, Y. H., Kao, T. W., Chen, R. J., Lai, J. L., & Perkasa, C. D. (2011). A novel intrusion detection system based on hierarchical clustering and support vector machines. *Expert Systems with Applications, 38*(1), 306–313. doi:10.1016/j.eswa.2010.06.066

Hsu, C. W., Chang, C. C., & Lin, C. J. (2003). *A practical guide to support vector classification.* Academic Press.

Iwendi, C., Uddin, M., Ansere, J. A., Nkurunziza, P., Anajemba, J. H., & Bashir, A. K. (2018). On Detection of Sybil Attack in Large-Scale VANETs Using Spider-Monkey Technique. *IEEE Access: Practical Innovations, Open Solutions*, 6, 47258–47267. doi:10.1109/ACCESS.2018.2864111

Jawandhiya, P. M., Ghonge, M. M., Ali, M. S., & Deshpande, J. S. (2010). A survey of mobile ad hoc network attacks. *International Journal of Engineering Science and Technology*, 2(9), 4063–4071.

Kabir, E., Hu, J., Wang, H., & Zhuo, G. (2018). A novel statistical technique for intrusion detection systems. *Future Generation Computer Systems*, 79, 303–318. doi:10.1016/j.future.2017.01.029

Kaelbling, L. P., Littman, M. L., & Moore, A. W. (1996). Reinforcement learning: A survey. *Journal of Artificial Intelligence Research*, 4, 237–285. doi:10.1613/jair.301

Kanungo, T., Mount, D. M., Netanyahu, N. S., Piatko, C. D., Silverman, R., & Wu, A. Y. (2002). An efficient k-means clustering algorithm: Analysis and implementation. *IEEE Transactions on Pattern Analysis and Machine Intelligence*, 24(7), 881–892. doi:10.1109/TPAMI.2002.1017616

Kaur, R., Sachdeva, M., & Kumar, G. (2016). Nature inspired feature selection approach for effective intrusion detection. *Indian Journal of Science and Technology*, 9(42). doi:10.17485/ijst/2016/v9i42/101555

Kevric, J., Jukic, S., & Subasi, A. (2017). An effective combining classifier approach using tree algorithms for network intrusion detection. *Neural Computing & Applications*, 28(1), 1051–1058. doi:10.100700521-016-2418-1

Khan, L., Awad, M., & Thuraisingham, B. (2007). A new intrusion detection system using support vector machines and hierarchical clustering. *The VLDB Journal*, 16(4), 507–521. doi:10.100700778-006-0002-5

Koc, L., Mazzuchi, T. A., & Sarkani, S. (2012). A network intrusion detection system based on a Hidden Naïve Bayes multiclass classifier. *Expert Systems with Applications*, 39(18), 13492–13500. doi:10.1016/j.eswa.2012.07.009

Kurt, B., Yıldız, Ç., Ceritli, T. Y., Sankur, B., & Cemgil, A. T. (2018). A Bayesian change point model for detecting SIP-based DDoS attacks. *Digital Signal Processing*, 77, 48–62. doi:10.1016/j.dsp.2017.10.009

Leung, K., & Leckie, C. (2005, January). Unsupervised anomaly detection in network intrusion detection using clusters. In *Proceedings of the Twenty-eighth Australasian conference on Computer Science-Volume 38* (pp. 333-342). Australian Computer Society, Inc.

Li, Y., Xia, J., Zhang, S., Yan, J., Ai, X., & Dai, K. (2012). An efficient intrusion detection system based on support vector machines and gradually feature removal method. *Expert Systems with Applications*, *39*(1), 424–430. doi:10.1016/j.eswa.2011.07.032

Liu, Z. (2011). A method of SVM with normalization in intrusion detection. *Procedia Environmental Sciences*, *11*, 256–262. doi:10.1016/j.proenv.2011.12.040

Long, N., & Thomas, R. (2001). Trends in denial of service attack technology. CERT Coordination Center, 648-651.

Lowd, D., & Domingos, P. (2005, August). Naive Bayes models for probability estimation. In *Proceedings of the 22nd international conference on Machine learning* (pp. 529-536). ACM.

Markey, J., & Atlasis, A. (2011). *Using decision tree analysis for intrusion detection: a how-to guide*. SANS Institute InfoSec Reading Room.

Meng, W., Li, W., Xiang, Y., & Choo, K. K. R. (2017). A bayesian inference-based detection mechanism to defend medical smartphone networks against insider attacks. *Journal of Network and Computer Applications*, *78*, 162–169. doi:10.1016/j.jnca.2016.11.012

Mohammed, M. N., & Sulaiman, N. (2012). Intrusion detection system based on SVM for WLAN. *Procedia Technology*, *1*, 313–317. doi:10.1016/j.protcy.2012.02.066

Müllner, D. (2011). *Modern hierarchical, agglomerative clustering algorithms*. arXiv preprint arXiv:1109.2378

Nguyen, D., Smith, N. A., & Rosé, C. P. (2011, June). Author age prediction from text using linear regression. In *Proceedings of the 5th ACL-HLT Workshop on Language Technology for Cultural Heritage, Social Sciences, and Humanities* (pp. 115-123). Association for Computational Linguistics.

Nguyen, D. H., & Widrow, B. (1990). Neural networks for self-learning control systems. *IEEE Control Systems Magazine*, *10*(3), 18–23. doi:10.1109/37.55119

Oyelade, O. J., Oladipupo, O. O., & Obagbuwa, I. C. (2010). *Application of k Means Clustering algorithm for prediction of Students Academic Performance.* arXiv preprint arXiv:1002.2425

Peddabachigari, S., Abraham, A., & Thomas, J. (2004). Intrusion detection systems using decision trees and support vector machines. *International Journal of Applied Science and Computations, USA, 11*(3), 118–134.

Peng, C. Y. J., Lee, K. L., & Ingersoll, G. M. (2002). An introduction to logistic regression analysis and reporting. *The Journal of Educational Research, 96*(1), 3–14. doi:10.1080/00220670209598786

Puthran, S., & Shah, K. (2016, September). Intrusion detection using improved decision tree algorithm with binary and quad split. In *International Symposium on Security in Computing and Communication* (pp. 427-438). Springer. 10.1007/978-981-10-2738-3_37

Rai, K., Devi, M. S., & Guleria, A. (2016). Decision Tree Based Algorithm for Intrusion Detection. *International Journal of Advanced Networking and Applications, 7*(4), 2828.

Rajasegarar, S., Leckie, C., & Palaniswami, M. (2008). Anomaly detection in wireless sensor networks. *IEEE Wireless Communications, 15*(4), 34–40. doi:10.1109/MWC.2008.4599219

Rmayti, M., Khatoun, R., Begriche, Y., Khoukhi, L., & Gaiti, D. (2017). A stochastic approach for packet dropping attacks detection in mobile Ad hoc networks. *Computer Networks, 121*, 53–64. doi:10.1016/j.comnet.2017.04.027

Rupareliya, J., Vithlani, S., & Gohel, C. (2016). Securing VANET by preventing attacker node using watchdog and Bayesian network theory. *Procedia Computer Science, 79*, 649–656. doi:10.1016/j.procs.2016.03.082

Salem, M. B., & Stolfo, S. J. (2009). *Masquerade attack detection using a search-behavior modeling approach.* Columbia University, Computer Science Department, Technical Report CUCS-027-09.

Santoro, D., Escudero-Andreu, G., Kyriakopoulos, K. G., Aparicio-Navarro, F. J., Parish, D. J., & Vadursi, M. (2017). A hybrid intrusion detection system for virtual jamming attacks on wireless networks. *Measurement, 109*, 79–87. doi:10.1016/j.measurement.2017.05.034

Schuba, C. L., Krsul, I. V., Kuhn, M. G., Spafford, E. H., Sundaram, A., & Zamboni, D. Analysis of a denial of service attack on TCP. In *Security and Privacy, 1997. Proceedings., 1997 IEEE Symposium on* (pp. 208-223). IEEE. 10.1109/SECPRI.1997.601338

Seber, G. A., & Lee, A. J. (2012). *Linear regression analysis* (Vol. 329). John Wiley & Sons.

Sengupta, N., Sen, J., Sil, J., & Saha, M. (2013). Designing of on line intrusion detection system using rough set theory and Q-learning algorithm. *Neurocomputing, 111*, 161–168. doi:10.1016/j.neucom.2012.12.023

Shah, S. A. R., & Issac, B. (2018). Performance comparison of intrusion detection systems and application of machine learning to Snort system. *Future Generation Computer Systems, 80*, 157–170. doi:10.1016/j.future.2017.10.016

Shamshirband, S., Amini, A., Anuar, N. B., Kiah, M. L. M., Teh, Y. W., & Furnell, S. (2014). D-FICCA: A density-based fuzzy imperialist competitive clustering algorithm for intrusion detection in wireless sensor networks. *Measurement, 55*, 212–226. doi:10.1016/j.measurement.2014.04.034

Shamshirband, S., Patel, A., Anuar, N. B., Kiah, M. L. M., & Abraham, A. (2014). Cooperative game theoretic approach using fuzzy Q-learning for detecting and preventing intrusions in wireless sensor networks. *Engineering Applications of Artificial Intelligence, 32*, 228–241. doi:10.1016/j.engappai.2014.02.001

Shiu, Y. S., Chang, S. Y., Wu, H. C., Huang, S. C. H., & Chen, H. H. (2011). Physical layer security in wireless networks: A tutorial. *IEEE Wireless Communications, 18*(2), 66–74. doi:10.1109/MWC.2011.5751298

Siddiqui, I. F., Qureshi, N. M. F., Shaikh, M. A., Chowdhry, B. S., Abbas, A., Bashir, A. K., & Lee, S. U. J. (n.d.). Stuck-at Fault Analytics of IoT Devices Using Knowledge-based Data Processing Strategy in Smart Grid. *Wireless Personal Communications*, 1-15.

Song, X., Wu, M., Jermaine, C., & Ranka, S. (2007). Conditional anomaly detection. *IEEE Transactions on Knowledge and Data Engineering, 19*(5), 631–645. doi:10.1109/TKDE.2007.1009

Sperandei, S. (2014). Understanding logistic regression analysis. *Biochemia medica. Biochemia Medica, 24*(1), 12–18. doi:10.11613/BM.2014.003 PMID:24627710

Stolfo, S. J., Salem, M. B., & Keromytis, A. D. (2012, May). Fog computing: Mitigating insider data theft attacks in the cloud. In *Security and Privacy Workshops (SPW), 2012 IEEE Symposium on* (pp. 125-128). IEEE.

Sutton, R. S., & Barto, A. G. (2018). *Reinforcement learning: An introduction.* MIT Press.

Tao, R., Yang, L., Peng, L., Li, B., & Cemerlic, A. (2009, March). A case study: Using architectural features to improve sophisticated denial-of-service attack detections. In *Computational Intelligence in Cyber Security, 2009. CICS'09. IEEE Symposium on* (pp. 13-18). IEEE.

Tapiador, J. E., & Clark, J. A. (2011). Masquerade mimicry attack detection: A randomised approach. *Computers & Security, 30*(5), 297-310.

Verma, A., & Ranga, V. (2018). Statistical analysis of CIDDS-001 dataset for Network Intrusion Detection Systems using Distance-based Machine Learning. *Procedia Computer Science, 125,* 709–716. doi:10.1016/j.procs.2017.12.091

Wang, Y. (2005). A multinomial logistic regression modeling approach for anomaly intrusion detection. *Computers & Security, 24*(8), 662–674. doi:10.1016/j.cose.2005.05.003

Zarpelao, B. B., Miani, R. S., Kawakani, C. T., & de Alvarenga, S. C. (2017). A survey of intrusion detection in Internet of Things. *Journal of Network and Computer Applications, 84,* 25–37. doi:10.1016/j.jnca.2017.02.009

APPENDIX

Table 2. List of Acronyms

SaaS	Software-as-a-service
ICT	Information and Communication Technology
IoT	Internet of Things
DoS	Denial of Service
TCP	Transmission Control Protocol
UDP	User Datagram Protocol
IDS	Intrusion Detection System
MDS	Misuse Detection System
ITU-T	International Telecommunications Union-Telecommunication
MBIDS	Misuse-Based Intrusion Detection System
ADS	Anomaly Detection System
NBTree	Naïve Bayes
CAD	Conditional Anomaly Detection
GMM	Gaussian Mixture Model
SIP	Session Initiation Protocol
DDoS	Distributed Denial of Service
MSN	Mobile Smartphone Networks
HNB	Hidden Naïve Bayes
SVM	Support Vector Machine
CT-SVM	Clustering Tree based on Support Vector Machine
NIDS	Network Intrusion Detection Systems
CIDDS	Coburg Network Intrusion detection Dataset
Weka	Waikato Environment for Knowledge Analysis
D-FICCA	Density-based Fuzzy imperialist competitive clustering algorithm
OLS	Ordinary Least Square
GBT	Gradient Boosting trees
BBS	Backside bus bandwidth
FSB	Front-side bus bandwidth
(CMAC)	Cerebellar Model Articulation Controller
G-FQL	Game based Fuzzy Q-learning
LEACH	Low energy Adaptive Clustering Hierarchy

Compilation of References

45. CFR parts 160 and 164: Modifications to the HIPAA Privacy, Security, Enforcement and Breach Notification Rules Under the HITECH Act and the Genetic Information Nondiscrimination Act: Other Modifications to the HIPAA Rules: Final Rule. (2013). The United States Health and Human Services. Retrieved from http://gpo.gov/fdsys/pkg/FR-2013-01-25/pdf/2013-01073.pdf

Abdi, F., & Haghighat, A. T. (2014). A hybrid rssi based localization algorithm dor wsn using a mobile anchor node. *International Conference on Computing, Communication and Networking Technologies.*

Abukeshipa & Barhoom. (2014). *Implementing and Comprising of OTP Techniques (TOTP, HOTP, CROTP) to Prevent Replay Attack in RADIUS Protocol.* Academic Press.

Acharya, S., Coats, B., Saluja, A., & Fuller, D. (2014). From Regulations to Practice: Achieving Information Security Compliance in Healthcare. *Proceedings of the 2014 Human Computer Interaction International Conference.*

Acharya, S., Coats, B., Saluja, A., & Fuller, D. (2013). A Roadmap for Information Security Assessment for Meaningful Use. *Proceedings of the 2013 IEEE/ACM International Symposium on Network Analysis and Mining for Health Informatics, Biomedicine and Bioinformatics.*

Acharya, S., Coats, B., Saluja, A., & Fuller, D. (2013). Secure Electronic Health Record Exchange: Achieving the Meaningful Use Objectives. *Proceedings of the 46th Hawaii International Conference on System Sciences*, 46, pp. 253-262. 10.1109/HICSS.2013.473

Adnan, M. A., Razzaque, M. A., Ahmed, I., & Isnin, I. F. (2014). Bio-Mimic optimization strategies in wireless sensor networks: A survey. *Sensors (Basel)*, *141*, 299–345. PMID:24368702

Ahmad, I., Basheri, M., Iqbal, M. J., & Rahim, A. (2018). Performance Comparison of Support Vector Machine, Random Forest, and Extreme Learning Machine for Intrusion Detection. *IEEE Access: Practical Innovations, Open Solutions, 6*, 33789–33795. doi:10.1109/ACCESS.2018.2841987

Ahmadon, M., Yamaguchi, S., Saon, S., & Mahamad, A. (2017). On Service Security Analysis for Event Log of IoT System Based on Data Petri Net. *2017 IEEE International Symposium on Consumer Electronics (ISCE).* 10.1109/ISCE.2017.8355531

Alrajeh, N. A., Bashir, M., & Shams, B. (2013). Localization techniques in wireless sensor networks". *International Journal of Distributed Sensor Networks, 96*, 1–9.

Alsheikh, M. A., Lin, S., Niyato, D., & Tan, H. (2014). Machine Learning in Wireless Sensor Networks: Algorithms, Strategies, and Applications. IEEE Communications Surveys & Tutorials, 16(4), 1996-2018. doi:10.1109/COMST.2014.2320099

Ambika, P. (2018). Machine Learning. In P. Raj & A. Raman (Eds.), *Handbook of Research on Cloud and Fog Computing Infrastructures for Data Science* (pp. 209–230). Hershey, PA: IGI Global. doi:10.4018/978-1-5225-5972-6.ch011

An Introductory Resource Guide for Implementing the Health Insurance Portability and Accountability Act (HIPAA) Security Rule. (2008). National Institute of Standards and Technologies. Retrieved from http://csrc.nist.gov/publications/nistpubs/800-66-Rev1/SP-800-66-Revision1.pdf

Anagnostopoulos, C. (2018). Weakly Supervised Learning: How to Engineer Labels for Machine Learning in Cyber-Security. *Data Science for Cyber-Security*, 195–226. doi:10.1142/9781786345646_010

Annas, G. (2003). HIPAA Regulations - A New Era of Medical-Record Privacy? *The New England Journal of Medicine, 348*(15), 1486–1490. doi:10.1056/NEJMlim035027 PMID:12686707

Appari, A., Anthony, D. L., & Johnson, M. E. (2009). *HIPAA Compliance: An Examination of Institutional and Market Forces.* Healthcare Information Management Systems Society.

Aral, S., & Weill, P. (2007). IT assets, organizational capabilities, and firm performance: How resource allocations and organizational differences explain performance variation. *Organization Science, 18*(5), 763–780. doi:10.1287/orsc.1070.0306

Arul, R., Raja, G., Bashir, A. K., Chaudry, J., & Ali, A. (2018). A Console GRID Leveraged Authentication and Key Agreement Mechanism for LTE/SAE. *IEEE Transactions on Industrial Informatics, 14*(6), 2677–2689. doi:10.1109/TII.2018.2817028

Arul, R., Raja, G., Kottursamy, K., Sathiyanarayanan, P., & Venkatraman, S. (2017). User path prediction based key caching and authentication mechanism for broadband wireless networks. *Wireless Personal Communications, 94*(4), 2645–2664. doi:10.100711277-016-3877-5

Atick, J. J., Griffin, P. A., & Norman Redlich, A. (2000). *Continuous video monitoring using face recognition for access control.* U.S. Patent 6,111,517.

Avital, A. (2017). *Authentication using facial recognition.* U.S. Patent 9,547,763.

Balcan, M. F., Liang, Y., & Gupta, P. (2014). Robust hierarchical clustering. *Journal of Machine Learning Research, 15*(1), 3831–3871.

Banerjee, S. P., & Woodard, D. L. (2012). Biometric authentication and identification using keystroke dynamics: A survey. *Journal of Pattern Recognition Research, 7*(1), 116–139. doi:10.13176/11.427

Behbood, V., Lu, J., & Zhang, G. (2014). Fuzzy Refinement Domain Adaptation for Long Term Prediction in Banking Ecosystem. *IEEE Transactions on Industrial Informatics, 10*(2), 1637 – 1646. 10.1109/TII.2012.2232935

Belyanina, L. A. (2018). Formation of an Effective Multi-Functional Model of the Research Competence of Students. In V. Mkrttchian & L. Belyanina (Eds.), *Handbook of Research on Students' Research Competence in Modern Educational Contexts* (pp. 17–39). Hershey, PA: IGI Global. doi:10.4018/978-1-5225-3485-3.ch002

Bernick, J. P. (2007). Very Fast Intrusion Detection by Multivariate Linear Regression. *Computer.*

Bharadwaj, S., Bharadwaj, A., & Bendoly, E. (2007). The Performance Effects of Complementarities Between Information Systems, Marketing, Manufacturing, and Supply Chain Processes. *Information Systems Research, 18*(4), 437–453. doi:10.1287/isre.1070.0148

Bhatti, B. M., & Sami, N. (2015). Building adaptive defense against cybercrimes using real-time data mining. *2015 First International Conference on Anti-Cybercrime (ICACC).* 10.1109/Anti-Cybercrime.2015.7351949

Bitam, S., Zeadally, S., & Mellouk, A. (2016). Bio-inspired cybersecurity for wireless sensor networks. *IEEE Communications Magazine*, *54*(6), 68–74. doi:10.1109/MCOM.2016.7497769

Blanzieri, E., & Bryl, A. (2008). A survey of learning-based techniques of email spam filtering. *Artificial Intelligence Review*, *29*(1), 63–92. doi:10.100710462-009-9109-6

Blumenthal, D., & Tavenner, M. (2010). The Meaningful Use Regulation for Electronic Health Records. *The New England Journal of Medicine*, *363*(6), 501–504. doi:10.1056/NEJMp1006114 PMID:20647183

Bolle, R. M., Nunes, S. L., Pankanti, S., Ratha, N. K., Smith, B. A., & Zimmerman, T. G. (2004). *Method for biometric-based authentication in wireless communication for access control.* U.S. Patent 6,819,219.

Boukerche, A., Oliveira, H., Nakamura, E., & Loureiro, A. (2007). Localization systems for wireless sensor networks. *IEEE Wireless Communications*, *14*(6), 6–12. doi:10.1109/MWC.2007.4407221

Bowyer, K. W., Chang, K., & Flynn, P. (2006). A survey of approaches and challenges in 3D and multi-modal 3D+ 2D face recognition. *Computer Vision and Image Understanding*, *101*(1), 1–15. doi:10.1016/j.cviu.2005.05.005

Breach Notification Rule. (2009). United States. Department of Health and Human Services. Office for Civil Rights. Retrieved from http://www.hhs.gov/ocr/privacy/hipaa/administrative/breachnotificationrule/index.html

Britz, D., & Miller, R. R. (2012). *Method and apparatus for eye-scan authentication using a liquid lens.* U.S. Patent 8,233,673.

Buczak, A. L., & Guven, E. (2016). A Survey of Data Mining and Machine Learning Methods for Cyber Security Intrusion Detection. *IEEE Communications Surveys and Tutorials*, *18*(2), 1153–1176. doi:10.1109/COMST.2015.2494502

Burmester, M., & de Medeiros, B. (2008). On the Security of Route Discovery in MANETs. *IEEE Transactions on Mobile Computing*, *8*(9), 1180–1188.

Bychkov, E. (2012). *Extended one-time password method and apparatus.* U.S. Patent 8,132, 243.

Ca͂nedo, J., & Skjellum, A. (2016). Using machine learning to secure IoT systems. *IEEE Conference Publications*, 219 – 222. 10.1109/PST.2016.7906930

Cadez, I. V., Smyth, P., & Mannila, H. (2001). Probabilistic modeling of transaction data with applications to profiling, visualization, and prediction. *Proceedings of the Seventh ACM SIGKDD International Conference on Knowledge Discovery and Data Mining - KDD '01*. doi:10.1145/502512.502523

Calvão, F. (2018). Crypto-miners: Digital labor and the power of blockchain technology. *Economic Anthropology*, 6(1), 123–134. doi:10.1002ea2.12136

Cannady, J. (2000, October). Next generation intrusion detection: Autonomous reinforcement learning of network attacks. In *Proceedings of the 23rd national information systems security conference* (pp. 1-12). Academic Press.

Carvalho. (2009). Security in Mobile Ad Hoc Networks. IEEE Security and Privacy, 6(2), 72–75.

Cayoglu, U. (2014). *Report: The Process Model Matching Contest 2013. In BPM 2013: Business Process Management Workshops* (pp. 442–463). Springer.

Chang, J., Tsou, P., Woungang, I., Chao, H., & Lai, C. (2015). Defending Against Collaborative Attacks by Malicious Nodes in MANETs: A Cooperative Bait Detection Approach. *IEEE Systems Journal*, 9(1), 65–75. doi:10.1109/JSYST.2013.2296197

Chang, R. K. (2002). Defending against flooding-based distributed denial-of-service attacks: A tutorial. *IEEE Communications Magazine*, 40(10), 42–51. doi:10.1109/MCOM.2002.1039856

Chaturvedi, S., Mishra, V., & Mishra, N. (2017). Sentiment analysis using machine learning for business intelligence. *IEEE International Conference on Power, Control, Signals and Instrumentation Engineering (ICPCSI)*, 2162-2166. 10.1109/ICPCSI.2017.8392100

Chen, Shi, Lee, Li, & Liu. (2014). The Customer Lifetime Value Prediction in Mobile Telecommunications. *IEEE Conference Publications*, 565 – 569.

Chen, M.-F., & Lien, G.-Y. (2008). The Mediating Role of Job Stress in Predicting Retail Banking Employees' Turnover Intentions in Taiwan. *IEED Conference Publications*, 393 - 398.

Chen, X., Weng, J., Lu, W., Xu, J., & Weng, J. (2018). Deep Manifold Learning Combined With Convolutional Neural Networks for Action Recognition. *IEEE Transactions on Neural Networks and Learning Systems*, 29(9), 3938–3952. doi:10.1109/TNNLS.2017.2740318 PMID:28922128

Chen, Y., Dass, S. C., & Jain, A. K. (2005). Fingerprint quality indices for predicting authentication performance. In *International Conference on Audio-and Video-Based Biometric Person Authentication* (pp. 160-170). Springer. 10.1007/11527923_17

Chiasson, Stobert, Forget, Biddle, & Van Oorschot. (2012). Persuasive cued click-points: Design, implementation, and evaluation of a knowledge-based authentication mechanism. *IEEE Transactions on Dependable and Secure Computing, 9*(2), 222-235.

Chuvakin, A., Schmidt, K., & Philips, C. (2013). *Logging and Log Management: The Authoritative Guide to Understanding the Concepts Surrounding Logging and Log Management*. Waltham, MA: Elsevier.

CMS EHR Meaningful Use Overview. (2012). Center for Medicare and Medicaid Services. Retrieved from https://www.cms.gov/Regulations-and-Guidance/Legislation/EHRIncentivePrograms/Meaningful_Use.html

Coats, B., Acharya, S., Saluja, A., & Fuller, D. (2012). HIPAA Compliance: How Do We Get There? A Standardized Framework for Enabling Healthcare Information Security & Privacy. *Proceedings of the 16th Colloquium for Information Systems Security Education*.

Conklin, D., & Walz. (2004). Password-based authentication: a system perspective. In *System Sciences*. Proceedings of the 37th Annual Hawaii International Conference on, 10.

E. Corchado, M. G. Romay, & A. M. Savio (Eds.). (2010). Hybrid Artificial Intelligent Systems, Part II*: 5th International Conference, HAIS 2010, San Sebastian, Spain*, June 23-25, 2010, *Proceedings* (Vol. 6076). Springer Science & Business Media.

Cortes, C., & Vapnik, V. (1995). Support-vector networks. *Machine Learning, 20*(3), 273–297. doi:10.1007/BF00994018

Cortopassi, M., & Endejan, E. (2013). *Method and apparatus for using pressure information for improved computer controlled handwriting recognition data entry and user authentication*. U.S. Patent 8,488,885.

Curry, S. M., Loomis, D. W., & Fox, C. W. (2000). *Method, apparatus, system, and firmware for secure transactions*. U.S. Patent 6,105,013.

Data and Reports. (2012). Center for Medicare and Medicaid Services. Retrieved from http://www.webcitation.org/6EMwIm36I

Data Pre-Processing. (2018). In *Wikipedia*. Retrieved from https://en.wikipedia.org/wiki/Data_pre-processing

David, S., Martikainen, J., Dudoit, S., & Ovaska, S. J. (2008). A general framework for statistical performance comparison of evolutionary computation algorithms. *Information Sciences, 17814*, 2870–2879.

Dawson, M., Omar, M., & Abramson, J. (n.d.). Understanding the Methods behind Cyber Terrorism. *Encyclopedia of Information Science and Technology, 3*, 1539–1549. doi:10.4018/978-1-4666-5888-2.ch147

Debnath, B., Solaimani, M., Gulzar, M., Arora, N., Lumezanu, C., Xu, J., . . . Khan, L. (2018). LogLens: A Real-Time Log Analysis System. *IEEE 38th International Conference on Distributed Computing Systems (ICDCS)*.

Deep Learning A-Z - ANN dataset. (n.d.). Retrieved from https://www.kaggle.com/filippoo/deep-learning-az-ann)

Denning, D. E. (1987). An intrusion-detection model. *IEEE Transactions on Software Engineering, SE-13*(2), 222–232. doi:10.1109/TSE.1987.232894

Dhurandher, Obaidat, & Verma, Gupta, & Dhurandher. (2016). FACES: Friend-Based Ad Hoc Routing Using Challenges to Establish Security in MANETs Systems. *IEEE Systems Journal, 5*(2), 176–188.

Dinur, I., Dolev, S., & Lodha, S. (Eds.). (2018). Lecture Notes in Computer Science Cyber Security Cryptography and Machine Learning. Springer. doi:10.1007/978-3-319-94147-9

Diro, A., & Chilamkurti, N. (2018). Leveraging LSTM Networks for Attack Detection in Fog-to-Things Communications. IEEE Communications Magazine, 56(9), 124-130. doi:10.1109/MCOM.2018.1701270

Doherty, L., Pister, K. S. J., & Ghaoui, L. E. (2001). Convex position estimation in wireless sensor networks. *Twentieth Annual Joint Conference of the IEEE Computer and Communications Societies Proceedings, 3*, 1655–1663.

Duffy, J. (1975). IFToMM symposium—Dublin, September 1974. *Mechanism and Machine Theory, 10*(2-3), 269. doi:10.1016/0094-114X(75)90030-0

Du, M., Li, F., Zheng, G., & Srikumar, V. (2017). DeepLog: Anomaly Detection and Diagnosis from System Logs through Deep Learning. In *Proceedings of the 2017 ACM SIGSAC Conference on Computer and Communications Security* (pp. 1285-1298). Dallas, TX: ACM. 10.1145/3133956.3134015

Dumas, M., La Rosa, M., Mendling, J., & Reijers, H. (2013). *Fundamentals of Business Process Management*. Springer. doi:10.1007/978-3-642-33143-5

Edgar, T. W., & Manz, D. O. (2017). Machine Learning. *Research Methods for Cyber Security*, 153–173. doi:10.1016/b978-0-12-805349-2.00006-6

EMR Adoption Trends. (2014). *HIMSS Analytics*. Retrieved from http://www.himssanalytics.org/stagesGraph.asp

Enforcement Results per Year. (2010). Center for Medicare and Medicaid Services. Retrieved from http://www.hhs.gov/ocr/privacy/hipaa/enforcement/data/historicalnumbers.html

Epler, P. (2013). Using the Response to Intervention (RtI) Service Delivery Model in Middle and High Schools. *International Journal for Cross-Disciplinary Subjects in Education*, 4(1), 1089–1098. doi:10.20533/ijcdse.2042.6364.2013.0154

Euzenat, J., & Shvaiko, P. (2013). Ontology Matching: State of the Art and Future Challenges. *IEEE Transactions on Knowledge and Data Engineering*, 25(1), 158–176. doi:10.1109/TKDE.2011.253

Feng, C., Wu, S., & Liu, N. (2017). A user-centric machine learning framework for cyber security operations center. *IEEE International Conference on Intelligence and Security Informatics (ISI)*, 173-175. 10.1109/ISI.2017.8004902

Feng, Y., Akiyama, H., Lu, L., & Sakurai, K. (2018). Feature Selection For Machine Learning-Based Early Detection of Distributed Cyber Attacks. *IEEE Conference Publications*, 173–180. 10.1109/DASC/PiCom/DataCom/CyberSciTec.2018.00040

Fichman, R., Kohli, R., & Krishnan, R. (2011). The Role of Information Systems in Healthcare: Current Research and Future Trends. *Information Systems Research*, 22(3), 419–428. doi:10.1287/isre.1110.0382

Fox, A., & Gribble, S. D. (1996). Security on the move: indirect authentication using Kerberos. *Proceedings of the 2nd annual international conference on Mobile computing and networking,* 155-164. 10.1145/236387.236439

Fuller, J. D., Ramsey, B. W., Rice, M. J., & Pecarina, J. M. (2017). Misuse-based detection of Z-Wave network attacks. *Computers & Security, 64,* 44–58. doi:10.1016/j.cose.2016.10.003

Gama, J., & de Carvalho, A. C. (2012). Machine Learning. In I. Management Association (Ed.), Machine Learning: Concepts, Methodologies, Tools and Applications (pp. 13-22). Hershey, PA: IGI Global. doi:10.4018/978-1-60960-818-7.ch102

Gangan, S. (2015). *A review of man-in-the-middle attacks.* arXiv preprint arXiv:1504.02115

Gardiner, J., & Nagaraja, S. (2016). On the Security of Machine Learning in Malware C8C Detection. *ACM Computing Surveys, 49*(3), 1–39. doi:10.1145/3003816

Gates, C., McNutt, J. J., Kadane, J. B., & Kellner, M. I. (2006, June). Scan detection on very large networks using logistic regression modeling. In *Computers and Communications, 2006. ISCC'06. Proceedings. 11th IEEE Symposium on* (pp. 402-408). IEEE. 10.1109/ISCC.2006.142

Gautam, S. K., & Om, H. (2015). Multivariate linear regression model for host based intrusion detection. In *Computational Intelligence in Data Mining-Volume 3* (pp. 361–371). New Delhi: Springer. doi:10.1007/978-81-322-2202-6_33

Ghosh & Datta. (2014). A Secure Addressing Scheme for Large-Scale Managed MANETs. *IEEE eTransactions on Network and Service Management, 12*(3), 483–495.

Giobbi, J. J., Brown, D. L., & Hirt, F. S. (2011). *Personal digital key differentiation for secure transactions.* U.S. Patent 7,904,718.

Goodfellow, I., Bengio, Y., & Courville, A. (2016). *Deep Learning.* MIT Press.

Gopakumar, A., & Jacob, L. (2008). Localization in wireless sensor networks using particle swarm optimization. *International Conference on Wireless, Mobile and Multimedia Networks,* 227–230.

Goyal, P., Batra, S., & Singh, A. (2010). A literature review of security attack in mobile ad-hoc networks. *International Journal of Computers and Applications, 9*(12), 11–15. doi:10.5120/1439-1947

Goyal, S., & Pattenh, M. (2015). Flower pollination algorithm based localization of wireless sensor network. *2nd International Conference on Recent Advances in Engineering and Computational Sciences*, Chandigarh, India. 10.1109/RAECS.2015.7453299

Goyal, S., & Patterh, M. S. (2013). Wireless sensor network localization based on BAT algorithm. *Int J Emerg Technol Comput Appl Sci, 5*, 507–512.

Gribaudo, M., Iacono, M., & Marrone, S. (2015). Exploiting Bayesian networks for the analysis of combined attack trees. *Electronic Notes in Theoretical Computer Science, 310*, 91–111. doi:10.1016/j.entcs.2014.12.014

Gu, Q., & Liu, P. (2007). Denial of service attacks. Handbook of Computer Networks: Distributed Networks, Network Planning, Control, Management, and New Trends and Applications, 3, 454-468. doi:10.1002/9781118256107.ch29

Guan, Y., Ghorbani, A. A., & Belacel, N. (2003, May). Y-means: A clustering method for intrusion detection. In *Electrical and Computer Engineering, 2003. IEEE CCECE 2003. Canadian Conference on* (Vol. 2, pp. 1083-1086). IEEE. 10.1109/CCECE.2003.1226084

Guan, Z., Bian, L., Shang, T., & Liu, J. (2018). When Machine Learning meets Security Issues: A survey. *IEEE Conference Publications*, 158 - 165. 10.1109/IISR.2018.8535799

Han, J., Kamber, M., & Pei, J. (2012). *Data Mining: Concepts and Techniques* (3rd ed.). Waltham, MA: Elsevier.

Harikrishnan, R., Jawahar, S. K., & Sridevi, P. (2014). Differential evolution approach for localization in wireless sensor networks. *IEEE International Conference on Computational Intelligence and Computing Research.* 10.1109/ICCIC.2014.7238536

He, Chan, & Guizani. (2017). Cyber Security Analysis and Protection of Wireless Sensor Networks for Smart Grid Monitoring. *IEEE Wireless Communications, 24*(6), 98-103. doi:10.1109/MWC.2017.1600283WC

He, S., Zhu, J., He, P., & Lyu, M. (2016). Experience Report: System Log Analysis for Anomaly Detection. *IEEE 27th International Symposium on Software Reliability Engineering (ISSRE).*

Health Reform in Action. (2010). United States White House. Retrieved from http://www.whitehouse.gov/healthreform/healthcare-overview

He, D., Liu, C., Quek, T. Q. S., & Wang, H. (2018). Transmit Antenna Selection in MIMO Wiretap Channels: A Machine Learning Approach. *IEEE Wireless Communications Letters, 7*(4), 634–637. doi:10.1109/LWC.2018.2805902

Hegadekatti, K. (2017). *Blockchain Technology - An Instrument of Economic Evolution?* SSRN Electronic Journal; doi:10.2139srn.2943960

Helms, M. M., Moore, R., & Ahmadi, M. (2008). Information Technology (IT) and the Healthcare Industry: A SWOT Analysis. *International Journal of Healthcare Information Systems and Informatics, 3*(1), 75–92. doi:10.4018/jhisi.2008010105

HER Incentive Programs. (2012). The Office of the National Coordinator for Health Information Technology. Retrieved from http://www.healthit.gov/providers-professionals/ehr-incentive-programs

HIPAA Administrative Simplification. (2006). United States. Department of Health and Human Services Office of Civil Rights. Retrieved from http://www.hhs.gov/ocr/privacy/hipaa/administrative/privacyrule/adminsimpregtext.pdf

HIPAA Compliance Review Analysis and Summary of Results. (2008). Center for Medicare and Medicaid Services. Retrieved from http://www.hhs.gov/ocr/privacy/hipaa/enforcement/cmscompliancerev08.pdf

Hirsch, R. D. (2013). Final HIPAA Omnibus Rule brings sweeping changes to health care privacy law: HIPAA privacy and security obligations extended to business associates and subcontractors. *Bloomberg Bureau of National Affairs Heath Law Reporter, 415*, 1–11.

Hoang, Q. T., Le, T. N., & Shin, Y. (2011). An RSS comparison based Localization in wireless sensor networks. *8th workshop on Positioning Navigation and communication WPNC*, 116–121. 10.1109/WPNC.2011.5961026

Hochreiter, S., & Schmidhuber, J. (1997). Long Short-Term Memory. *Neural Computation, 9*(8), 1735–1780. doi:10.1162/neco.1997.9.8.1735 PMID:9377276

Horng, S. J., Su, M. Y., Chen, Y. H., Kao, T. W., Chen, R. J., Lai, J. L., & Perkasa, C. D. (2011). A novel intrusion detection system based on hierarchical clustering and support vector machines. *Expert Systems with Applications, 38*(1), 306–313. doi:10.1016/j.eswa.2010.06.066

Hsu, C. W., Chang, C. C., & Lin, C. J. (2003). *A practical guide to support vector classification*. Academic Press.

Hunt, A. K., & Schalk, T. B. (1992). *Simultaneous speaker-independent voice recognition and verification over a telephone network.* U.S. Patent 5,127,043.

Investopedia. (2018). *Introduction to the Chinese Banking.* Retrieved from https://www.investopedia.com/articles/economics/11/chinese-banking-system.asp

Islam, N., Das, S., & Chen, Y. (2017). On-Device Mobile Phone Security Exploits Machine Learning. *IEEE Pervasive Computing, 16*(2), 92–96. doi:10.1109/MPRV.2017.26

Iwendi, C., Uddin, M., Ansere, J. A., Nkurunziza, P., Anajemba, J. H., & Bashir, A. K. (2018). On Detection of Sybil Attack in Large-Scale VANETs Using Spider-Monkey Technique. *IEEE Access: Practical Innovations, Open Solutions, 6,* 47258–47267. doi:10.1109/ACCESS.2018.2864111

Jawandhiya, P. M., Ghonge, M. M., Ali, M. S., & Deshpande, J. S. (2010). A survey of mobile ad hoc network attacks. *International Journal of Engineering Science and Technology, 2*(9), 4063–4071.

Jenab, K., Khoury, S., & LaFevor, K. (2018). Flow-Graph and Markovian Methods for Cyber Security Analysis. In I. Management Association (Ed.), Cyber Security and Threats: Concepts, Methodologies, Tools, and Applications (pp. 674-702). Hershey, PA: IGI Global. doi:10.4018/978-1-5225-5634-3.ch036

Jenab, K., Khoury, S., & LaFevor, K. (2016). Flow-Graph and Markovian Methods for Cyber Security Analysis. *International Journal of Enterprise Information Systems, 12*(1), 59–84. doi:10.4018/IJEIS.2016010104

Jiajia, L., Nei, K., Jianfeng, M., & Naoto, K. (2015). Device-to-device vommunication in LTE-advanced networks: A survey. *IEEE Communications Surveys and Tutorials, 17*(4), 1923–1940. doi:10.1109/COMST.2014.2375934

Johnson, M. E., Goetz, E., & Pfleeger, S. L. (2009). Security through Information Risk Management. *IEEE Security and Privacy, 7*(3), 45–52. doi:10.1109/MSP.2009.77

Jonesa & Teeling. (2006). The evolution of echolocation in bats. *Trends in Ecology and Evolution, 213,* 149–156.

Kabir, E., Hu, J., Wang, H., & Zhuo, G. (2018). A novel statistical technique for intrusion detection systems. *Future Generation Computer Systems, 79,* 303–318. doi:10.1016/j.future.2017.01.029

Kaelbling, L. P., Littman, M. L., & Moore, A. W. (1996). Reinforcement learning: A survey. *Journal of Artificial Intelligence Research, 4*, 237–285. doi:10.1613/jair.301

Kanevsky, D., & Maes, S. H. (2000). *Apparatus and methods for providing repetitive enrollment in a plurality of biometric recognition systems based on an initial enrollment.* U.S. Patent 6,092,192.

Kanungo, T., Mount, D. M., Netanyahu, N. S., Piatko, C. D., Silverman, R., & Wu, A. Y. (2002). An efficient k-means clustering algorithm: Analysis and implementation. *IEEE Transactions on Pattern Analysis and Machine Intelligence, 24*(7), 881–892. doi:10.1109/TPAMI.2002.1017616

Kaur, R., Sachdeva, M., & Kumar, G. (2016). Nature inspired feature selection approach for effective intrusion detection. *Indian Journal of Science and Technology, 9*(42). doi:10.17485/ijst/2016/v9i42/101555

Kayworth, T., & Whitten, D. (2010). Effective Information Security Requires a Balance of Social and Technology Factors. *MIS Quarterly Executive, 9*(3), 163–175.

Kevric, J., Jukic, S., & Subasi, A. (2017). An effective combining classifier approach using tree algorithms for network intrusion detection. *Neural Computing & Applications, 28*(1), 1051–1058. doi:10.100700521-016-2418-1

Khan, M. S. (Ed.). (2019). Machine Learning and Cognitive Science Applications in Cyber Security. Academic Press. doi:10.4018/978-1-5225-8100-0

Khan, L., Awad, M., & Thuraisingham, B. (2007). A new intrusion detection system using support vector machines and hierarchical clustering. *The VLDB Journal, 16*(4), 507–521. doi:10.100700778-006-0002-5

Khan, M. A., & Hasan, A. (2008). Pseudo-random number based authentication to counter denial of service attacks on 802.11.Wireless and Optical Communications Networks, 2008. In *WOCN'08. 5th IFIP International Conference.* (pp. 1-5). IEEE.

Khan, M. S., Ferens, K., & Kinsner, W. (2014). A Chaotic Complexity Measure for Cognitive Machine Classification of Cyber-Attacks on Computer Networks. *International Journal of Cognitive Informatics and Natural Intelligence, 8*(3), 45–69. doi:10.4018/IJCINI.2014070104

Koc, L., Mazzuchi, T. A., & Sarkani, S. (2012). A network intrusion detection system based on a Hidden Naïve Bayes multiclass classifier. *Expert Systems with Applications, 39*(18), 13492–13500. doi:10.1016/j.eswa.2012.07.009

Kravets, A., Shcherbakov, M., Kultsova, M., & Shabalina, O. (2015). *Creativity in Intelligent, Technologies and Data Science – 2015*. First Conference, CIT&DS 2015, Volgograd, Russia. doi: 10.1007/978-3-319-23766-4

Kulkarni, R., Venayagamoorthy, G., & Cheng, M. (2009). Bio-inspired node localization in wireless sensor networks. *International Conference on Systems, Man and Cybernetics*, 205–210. 10.1109/ICSMC.2009.5346107

Kumar, R., & Rituraj, S. (2017). Using Data Mining and Machine Learning Methods for Cyber Security Intrusion Detection. *International Journal of Recent Trends in Engineering and Research*, 3(4), 109–111. doi:10.23883/IJRTER.2017.3117.9NWQV

Kurt, B., Yıldız, Ç., Ceritli, T. Y., Sankur, B., & Cemgil, A. T. (2018). A Bayesian change point model for detecting SIP-based DDoS attacks. *Digital Signal Processing*, 77, 48–62. doi:10.1016/j.dsp.2017.10.009

Kussul, N., & Skakun, S. (2005). Intelligent System for Users' Activity Monitoring in Computer Networks. Intelligent Data Acquisition and Advanced Computing Systems: Technology and Applications Conference.

Kwon, J., & Johnson, M. E. (2013). Healthcare Security Strategies for Regulatory Compliance and Data Security. *Proceedings of the 46th Hawaii International Conference on System Sciences*. 10.1109/HICSS.2013.246

Layman, L., Diffo, S., & Zazworka, N. (2014). Human Factors in Webserver Log File Analysis: A Controlled Experiment on Investigating Malicious Activity. *Proceedings of the 2014 Symposium and Bootcamp on the Science of Security*. 10.1145/2600176.2600185

Lee, Park, Lim, Kim, & Jeong. (2014). Server authentication for blocking unapproved WOW access. *2014 International Conference on Big Data and Smart Computing (BIGCOMP)*, 155-159. 10.1109/BIGCOMP.2014.6741427

Leung, K., & Leckie, C. (2005, January). Unsupervised anomaly detection in network intrusion detection using clusters. In *Proceedings of the Twenty-eighth Australasian conference on Computer Science-Volume 38* (pp. 333-342). Australian Computer Society, Inc.

Li, C.-Y., Jiang, L., Liang, A.-N., & Liao, L.-J. (2005). A User-Centric Machine Learning Framework for Cyber Security Operations Center. *IEEE Conference Publications*, 173-175.

Lima, M., Zarpelao, B., Sampaio, L., Rodrigues, J., Abrao, T., & Proenca, M. (2010). Anomaly Detection Using Baseline and K-Means Clustering. *International Conference on Software, Telecommunications and Computer Networks (SoftCOM)*.

Liu, B. (2011). *Web Data Mining: Exploring Hyperlinks, Contents, and Usage Data* (2nd ed.). Chicago, IL: Springer. doi:10.1007/978-3-642-19460-3

Liu, Q., Li, P., Zhao, W., Cai, W., Yu, S., & Leung, V. C. M. (2018). A Survey on Security Threats and Defensive Techniques of Machine Learning: A Data Driven View. *IEEE Access: Practical Innovations, Open Solutions*, *6*, 12103–12117. doi:10.1109/ACCESS.2018.2805680

Liu, Z. (2011). A method of SVM with normalization in intrusion detection. *Procedia Environmental Sciences*, *11*, 256–262. doi:10.1016/j.proenv.2011.12.040

Li, Y., Xia, J., Zhang, S., Yan, J., Ai, X., & Dai, K. (2012). An efficient intrusion detection system based on support vector machines and gradually feature removal method. *Expert Systems with Applications*, *39*(1), 424–430. doi:10.1016/j.eswa.2011.07.032

Long, N., & Thomas, R. (2001). Trends in denial of service attack technology. CERT Coordination Center, 648-651.

Lowd, D., & Domingos, P. (2005, August). Naive Bayes models for probability estimation. In *Proceedings of the 22nd international conference on Machine learning* (pp. 529-536). ACM.

Luke & Taylor. (2015). *Apparatus, method, and article for authentication, security and control of power storage devices, such as batteries*. U.S. Patent 9,182,244.

Luo, Y., & Tsai, J. J. P. (2008). A Framework for Extrusion Detection Using Machine Learning. *IEEE Conference Publications*, 83 – 88.

M'Raihi, D., Bellare, M., Hoornaert, F., Naccache, D., & Ranen, O. (2005). *An Hmac-based One-Time Password algorithm*. No. RFC 4226.

Mahaffey, Richardson, Salomon, Croy, Walker, Buck, … Golombek. (2016). *Multi-factor authentication and comprehensive login system for client-server networks*. U.S. Patent 9,374,369.

Malaiya, R., Kwon, D., Kim, J., Suh, S., Kim, H., & Kim, I. (2018). An Empirical Evaluation of Deep Learning for Network Anomaly Detection. *International Conference on Computing, Networking and Communications (ICNC)*. 10.1109/ICCNC.2018.8390278

Mangialardo & Duarte. (2015). Integrating Static and Dynamic Malware Analysis Using Machine Learning. *IEEE Latin America Transactions, 13*(9), 3080-3087.

Mao, G., Fidan, B., & Anderson, B. D. (2007). Wireless sensor network localization techniques ". *Computer Networks, 51*(10), 2529–2553. doi:10.1016/j.comnet.2006.11.018

Markey, J., & Atlasis, A. (2011). *Using decision tree analysis for intrusion detection: a how-to guide.* SANS Institute InfoSec Reading Room.

McDaniel, P., Papernot, N., & Celik, Z. B. (2016). Machine Learning in Adversarial Settings. *IEEE Security and Privacy, 14*(3), 68–72. doi:10.1109/MSP.2016.51

Mendling, J. (2018). Blockchains for Business Process Management – Challenges and Opportunities. ACM Trans. Manag. Inform. Syst., 9.

Meng, W., Li, W., Xiang, Y., & Choo, K. K. R. (2017). A bayesian inference-based detection mechanism to defend medical smartphone networks against insider attacks. *Journal of Network and Computer Applications, 78*, 162–169. doi:10.1016/j.jnca.2016.11.012

Meng, Y., Wong, D. S., & Schlegel, R. (2012). Touch gestures based biometric authentication scheme for touchscreen mobile phones. In *International Conference on Information Security and Cryptology* (pp. 331-350). Springer.

Meulen, R. (2017). *Gartner Says 8.4 Billion Connected "Things" Will Be in Use in 2017, Up 31 Percent From 2016.* Retrieved from https://www.gartner.com/en/newsroom/press-releases/2017-02-07-gartner-says-8-billion-connected-things-will-be-in-use-in-2017-up-31-percent-from-2016

Mihoubi, M., Rahmoun, A., & Lorenz, P. (2017). Metaheuristic RSSI based for node localization in distributed wireless sensor network. *Global Information Infrastructure and Networking Symposium (GIIS)*, 64-70. 10.1109/GIIS.2017.8169811

Miloud, M., Rahmoun, A., Lorenz, P., & Lasla, N. (2017). An effective Bat algorithm for node localization in distributed wireless sensor network. *Security and Privacy.* . doi:10.1002py2.7

Miloud, M., Abdellatif, R., & Lorenz, P. (2019). Moth Flame Optimization Algorithm Range-Based for Node Localization Challenge in Decentralized Wireless Sensor Network. *International Journal of Distributed Systems and Technologies, 10*(1), 82–109. doi:10.4018/IJDST.2019010106

Mizrah, L. L. (2011). *Two-channel challenge-response authentication method in random partial shared secret recognition system.* U.S. Patent 8,006,300.

Mkrttchian, V. (2015a), Use Online Multi-Cloud Platform Lab with Intellectual Agents: Avatars for Study of Knowledge Visualization & Probability Theory in Bioinformatics. International Journal of Knowledge Discovery in Bioinformatics, 5(1), 11-23. Doi:10.4018/IJKDB.2015010102

Mkrttchian, V. (2015b). Modeling using of Triple H-Avatar Technology in online Multi-Cloud Platform Lab. In M. Khosrow-Pour (Ed.), Encyclopedia of Information Science and Technology (3rd Ed.). (pp. 4162-4170). Hershey, PA: IGI Global. Doi:10.4018/978-1-4666-5888-2.ch409

Mkrttchian, V., Kataev, M., Hwang, W., Bedi, S., & Fedotova, A. (2016), Using Plug-Avatars "hhh" Technology Education as Service-Oriented Virtual Learning Environment in Sliding Mode. In Leadership and Personnel Management: Concepts, Methodologies, Tools, and Applications (pp. 890-902). IGI Global. Doi:10.4018/978-1-4666-9624-2.ch039

Mkrttchian, V., Kataev, M., Shih, T., Kumar, M., & Fedotova, A. (2014). Avatars "HHH" Technology Education Cloud Platform on Sliding Mode Based Plug-Ontology as a Gateway to Improvement of Feedback Control Online Society. International Journal of Information Communication Technologies and Human Development, 6(3), 13-31. Doi:10.4018/ijicthd.2014070102

Mkrttchian, V. (2011). Use 'hhh" technology in transformative models of online education. In G. Kurubacak & T. Vokan Yuzer (Eds.), *Handbook of research on transformative online education and liberation: Models for social equality* (pp. 340–351). Hershey, PA: IGI Global. doi:10.4018/978-1-60960-046-4.ch018

Mkrttchian, V. (2012). Avatar manager and student reflective conversations as the base for describing meta-communication model. In G. Kurubacak, T. Vokan Yuzer, & U. Demiray (Eds.), *Meta-communication for reflective online conversations: Models for distance education* (pp. 340–351). Hershey, PA: IGI Global. doi:10.4018/978-1-61350-071-2.ch005

Mkrttchian, V. (2017). Project-Based Learning for Students with Intellectual Disabilities. In P. L. Epler (Ed.), *Instructional Strategies in General Education and Putting the Individuals With Disabilities Act (IDEA) Into Practice* (pp. 196–221). Hershey, PA: IGI Global. doi:10.4018/978-1-5225-3111-1.ch007

Mkrttchian, V., & Aleshina, E. (2017). *Sliding Mode in Intellectual Control and Communication: Emerging Research and Opportunities*. Hershey, PA: IGI Global; doi:10.4018/978-1-5225-2292-8

Mkrttchian, V., & Aleshina, E. (2017). The Sliding Mode Technique and Technology (SM T&T) According to VardanMkrttchian in Intellectual Control(IC). In *Sliding Mode in Intellectual Control and Communication: Emerging Research and Opportunities* (pp. 1–9). Hershey, PA: IGI Global. doi:10.4018/978-1-5225-2292-8.ch001

Mkrttchian, V., & Belyanina, L. (Eds.). (2018). *Handbook of Research on Students' Research Competence in Modern Educational Contexts*. Hershey, PA: IGI Global. doi:10.4018/978-1-5225-3485-3

Mkrttchian, V., Bershadsky, A., Bozhday, A., & Fionova, L. (2015). Model in SM of DEE Based on Service Oriented Interactions at Dynamic Software Product Lines. In G. Eby & T. Vokan Yuzer (Eds.), *Identification, Evaluation, and Perceptions of Distance Education Experts* (pp. 230–247). Hershey, PA: IGI Global. doi:10.4018/978-1-4666-8119-4.ch014

Mkrttchian, V., Bershadsky, A., Bozhday, A., Kataev, M., & Kataev, S. (Eds.). (2016b). *Handbook of Research on Estimation and Control Techniques in E-Learning systems*. Hershey, PA: IGI Global. doi:10.4018/978-1-4666-9489-7

Mkrttchian, V., Bershadsky, A., Bozhday, A., Noskova, T., & Miminova, S. (2016a). Development of a Global Policy of All-Pervading E-Learning, Based on Transparency, Strategy, and Model of Cyber Triple H-Avatar. In G. Eby, T. V. Yuser, & S. Atay (Eds.), *Developing Successful Strategies for Global Policies and Cyber Transparency in E-Learning* (pp. 207–221). Hershey, PA: IGI Global. doi:10.4018/978-1-4666-8844-5.ch013

Mkrttchian, V., Kataev, M., Hwang, W., Bedi, S., & Fedotova, A. (2014). Using Plug-Avatars "hhh" Technology Education as Service-Oriented Virtual Learning Environment in Sliding Mode. In G. Eby & T. Vokan Yuzer (Eds.), *Emerging Priorities and Trends in Distance Education: Communication, Pedagogy, and Technology*. Hershey, PA: IGI Global. doi:10.4018/978-1-4666-5162-3.ch004

Mkrttchian, V., Palatkin, I., Gamidullaeva, L. A., & Panasenko, S. (2019). About Digital Avatars for Control Systems Using Big Data and Knowledge Sharing in Virtual Industries. In A. Gyamfi & I. Williams (Eds.), *Big Data and Knowledge Sharing in Virtual Organizations* (pp. 103–116). Hershey, PA: IGI Global. doi:10.4018/978-1-5225-7519-1.ch004

Mkrttchian, V., Veretekhina, S., Gavrilova, O., Ioffe, A., Markosyan, S., & Chernyshenko, S. V. (2019). The Cross-Cultural Analysis of Australia and Russia: Cultures, Small Businesses, and Crossing the Barriers. In U. Benna (Ed.), *Industrial and Urban Growth Policies at the Sub-National, National, and Global Levels* (pp. 229–249). Hershey, PA: IGI Global. doi:10.4018/978-1-5225-7625-9.ch012

Mohammed, M. N., & Sulaiman, N. (2012). Intrusion detection system based on SVM for WLAN. *Procedia Technology, 1*, 313–317. doi:10.1016/j.protcy.2012.02.066

Morabito, V. (2017a). Blockchain and Enterprise Systems. *Business Innovation Through Blockchain*, 125–142. doi:10.1007/978-3-319-48478-5_7

Morabito, V. (2017b). The Blockchain Paradigm Change Structure. *Business Innovation Through Blockchain*, 3–20. doi:10.1007/978-3-319-48478-5_1

Morabito, V. (2017c). Blockchain Practices. *Business Innovation through Blockchain*, 145–166. doi:10.1007/978-3-319-48478-5_8

Morales, A., Travieso, C. M., Ferrer, M. A., & Alonso, J. B. (2011). Improved finger-knuckle-print authentication based on orientation enhancement. *Electronics Letters, 47*(6), 380–381. doi:10.1049/el.2011.0156

Morelande, M., Moran, B., & Brazil, M. (2008). Bayesian node localisation in wireless sensor networks. *IEEE International Conference on Acoustics, Speech and Signal Processing*, 2545–2548. 10.1109/ICASSP.2008.4518167

Moussa, A., & El-sheimy, N. (2010). Localization of wireless sensor network using Bees optimization algorithm. *IEEE International Symposium on Signal Processing and Information Technology*, 478–481. 10.1109/ISSPIT.2010.5711760

Mozaffari-Kermani, M., Sur-Kolay, S., Raghunathan, A., & Jha, N. K. (2015). Systematic Poisoning Attacks on and Defenses for Machine Learning in Healthcare. *IEEE Journal of Biomedical and Health Informatics, 19*(6), 1893–1905. doi:10.1109/JBHI.2014.2344095 PMID:25095272

M'Raihi, Machani, Pei, & Rydell. (2011). *Top: Time-based one-time password algorithm*. No. RFC 6238.

M'Raihi, Machani, Pei, & Rydell. (2011). *TOTP: Time-based One-Time Password algorithm*. No. RFC 6238.

Müllner, D. (2011). *Modern hierarchical, agglomerative clustering algorithms*. arXiv preprint arXiv:1109.2378

Nguyen, D. H., & Widrow, B. (1990). Neural networks for self-learning control systems. *IEEE Control Systems Magazine, 10*(3), 18–23. doi:10.1109/37.55119

Nguyen, D. Q., Toulgoat, M., & Lamont, L. (2011). Impact of trust-based security association and mobility on the delay metric in MANET. *Journal of Communications and Networks (Seoul), 18*(1), 105–111.

Nguyen, D., Smith, N. A., & Rosé, C. P. (2011, June). Author age prediction from text using linear regression. In *Proceedings of the 5th ACL-HLT Workshop on Language Technology for Cultural Heritage, Social Sciences, and Humanities* (pp. 115-123). Association for Computational Linguistics.

Nichols, T. J., & Thompson, D. L. (2005). *User authentication in medical device systems.* U.S. Patent 6,961,448.

Olah, C. (2015). *Understanding LSTM Networks.* Retrieved from http://colah.github.io/posts/2015-08-Understanding-LSTMs/

Oyelade, O. J., Oladipupo, O. O., & Obagbuwa, I. C. (2010). *Application of k Means Clustering algorithm for prediction of Students Academic Performance.* arXiv preprint arXiv:1002.2425

Ozay, M., Esnaola, I., Yarman Vural, F. T., Kulkarni, S. R., & Poor, H. V. (2016). Machine Learning Methods for Attack Detection in the Smart Grid. *IEEE Transactions on Neural Networks and Learning Systems, 27*(8), 1773–1786. doi:10.1109/TNNLS.2015.2404803 PMID:25807571

Pal, A. (2010). Localization algorithms in wireless sensor networks: Current approaches and future challenges. *Netw Protoc Algorithm, 21*, 45–73.

Pat Research. (2018). *Artificial Intelligence Platforms.* Retrieved from https://www.predictiveanalyticstoday.com/artificial-intelligence-platforms/

Patwari, N., Ash, J. N., Kyperountas, S., Hero, A. O., Moses, R. L., & Correal, N. S. (2005). Locating the nodes: Cooperative localization in wireless sensor networks. *IEEE Signal Processing Magazine, 224*(4), 54–69. doi:10.1109/MSP.2005.1458287

Paul, P. S., & Dharwadkar, N. V. (2017). Analysis of Banking Data Using Machine Learning. *IEEE Conference Publications*, 876 - 881.

Peddabachigari, S., Abraham, A., & Thomas, J. (2004). Intrusion detection systems using decision trees and support vector machines. *International Journal of Applied Science and Computations, USA, 11*(3), 118–134.

Peng, C. Y. J., Lee, K. L., & Ingersoll, G. M. (2002). An introduction to logistic regression analysis and reporting. *The Journal of Educational Research*, *96*(1), 3–14. doi:10.1080/00220670209598786

Philip, T. D., & Soo, Y. S. (2016). Range Based Wireless Node Localization Using Dragonfly Algorithm. *Eighth International Conference on Ubiquitous and Future Networks (ICUFN)*. DOI: 10.1109/ICUFN.2016.7536950

Plotnikov, V., Vertakova, Y., & Leontyev, E. (2016). Evaluation of the effectiveness of the telecommunication company's cluster management. *Economic Computation and Economic Cybernetics Studies and Research*, *50*(4), 109–118.

Popp, M'raihi, & Hart. (2011). One-*time password*. U.S. Patent 8,087,074.

Powell, V. (n.d.). *Principal Component Analysis: Explained Visually*. Retrieved from http://setosa.io/ev/principal-component-analysis/

Puthran, S., & Shah, K. (2016, September). Intrusion detection using improved decision tree algorithm with binary and quad split. In *International Symposium on Security in Computing and Communication* (pp. 427-438). Springer. 10.1007/978-981-10-2738-3_37

Qureshi, S. A., Rehman, A. S., Qamar, A. M., Kamal, A., & Rehman, A. (2013). Customer Churn Prediction Modelling Based on Behavioural Patterns Analysis using Deep Learning. *IEEE Conference Publications*, 1 - 6.

Rabkin, A. (2008). Personal knowledge questions for fallback authentication: Security questions in the era of Facebook. *Proceedings of the 4th symposium on Usable privacy and security*, 13-23. 10.1145/1408664.1408667

Rai, K., Devi, M. S., & Guleria, A. (2016). Decision Tree Based Algorithm for Intrusion Detection. *International Journal of Advanced Networking and Applications*, *7*(4), 2828.

Rajakumar, Amudhavel, Dhavachelvan, & Vengattaraman. (2017). GWO-LPWSN: Grey Wolf Optimization Algorithm for Node Localization Problem in Wireless Sensor Networks. *Journal of Computer Networks and Communications*. . doi:10.1155/2017/7348141

Rajasegarar, S., Leckie, C., & Palaniswami, M. (2008). Anomaly detection in wireless sensor networks. *IEEE Wireless Communications*, *15*(4), 34–40. doi:10.1109/MWC.2008.4599219

Rassan & Shaher. (2013). Securing mobile cloud using fingerprint authentication. *International Journal of Network Security & Its Applications, 5*(6), 41.

Rath & Oreku. (2018). Security Issues in Mobile Devices and Mobile Adhoc Networks. In Mobile Technologies and Socio-Economic Development in Emerging Nations. IGI Global. doi:10.4018/978-1-5225-4029-8.ch009

Rath & Pattanayak. (2019). Security Protocol with IDS Framework Using Mobile Agent in Robotic MANET. *International Journal of Information Security and Privacy, 13*(1), 46-58. Doi:10.4018/IJISP.2019010104

Rath & Swain. (2018). IoT Security: A Challenge in Wireless Technology. *International Journal of Emerging Technology and Advanced Engineering, 8*(4), 43-46.

Rath, M. (2018). An Analytical Study of Security and Challenging Issues in Social Networking as an Emerging Connected Technology. *Proceedings of 3rd International Conference on Internet of Things and Connected Technologies (ICIoTCT)*.

Rath, M., & Panda, M. R. (2017). MAQ system development in mobile ad-hoc networks using mobile agents. *IEEE 2nd International Conference on Contemporary Computing and Informatics (IC3I)*, 794-798.

Rath, M., & Pati, B. (2018). Security Assertion of IoT Devices Using Cloud of Things Perception. International Journal of Interdisciplinary Telecommunications and Networking, 11(2).

Rath, M., & Pattanayak, B. (2018). Technological improvement in modern health care applications using Internet of Things (IoT) and proposal of novel health care approach. *International Journal of Human Rights in Healthcare.* doi:10.1108/IJHRH-01-2018-0007

Rath, M., & Pattanayak, B. K. (2018). Monitoring of QoS in MANET Based Real Time Applications. In Information and Communication Technology for Intelligent Systems Volume 2. ICTIS. Smart Innovation, Systems and Technologies (vol. 84, pp. 579-586). Springer. doi:10.1007/978-3-319-63645-0_64

Rath, M., Pati, B., & Pattanayak, B. (2019). Manifold Surveillance Issues in Wireless Network and the Secured Protocol. *International Journal of Information Security and Privacy, 13*(3).

Rath, M. (2017). Resource provision and QoS support with added security for client side applications in cloud computing. *International Journal of Information Technology, 9*(3), 1–8.

Rath, M., & Pati, B. (2017). *Load balanced routing scheme for MANETs with power and delay optimisation. International Journal of Communication Network and Distributed Systems* , 19.

Rath, M., Pati, B., Panigrahi, C. R., & Sarkar, J. L. (2019). QTM: A QoS Task Monitoring System for Mobile Ad hoc Networks. In P. Sa, S. Bakshi, I. Hatzilygeroudis, & M. Sahoo (Eds.), *Recent Findings in Intelligent Computing Techniques. Advances in Intelligent Systems and Computing* (Vol. 707). Singapore: Springer. doi:10.1007/978-981-10-8639-7_57

Rath, M., Pati, B., & Pattanayak, B. K. (2017). Cross layer based QoS platform for multimedia transmission in MANET. *11th International Conference on Intelligent Systems and Control (ISCO)*, 402-407. 10.1109/ISCO.2017.7856026

Rath, M., & Pattanayak, B. (2017). MAQ:A Mobile Agent Based QoS Platform for MANETs. *International Journal of Business Data Communications and Networking, IGI Global, 13*(1), 1–8. doi:10.4018/IJBDCN.2017010101

Rath, M., & Pattanayak, B. K. (2018). SCICS: A Soft Computing Based Intelligent Communication System in VANET. Smart Secure Systems – IoT and Analytics Perspective. *Communications in Computer and Information Science, 808*, 255–261. doi:10.1007/978-981-10-7635-0_19

Rath, M., Pattanayak, B. K., & Pati, B. (2017). *Energetic Routing Protocol Design for Real-time Transmission in Mobile Ad hoc Network. In Computing and Network Sustainability, Lecture Notes in Networks and Systems* (Vol. 12). Singapore: Springer.

Rath, M., Swain, J., Pati, B., & Pattanayak, B. K. (2018). *Attacks and Control in MANET. In Handbook of Research on Network Forensics and Analysis Techniques* (pp. 19–37). IGI Global.

Regulations and Guidance. (2004). Center for Medicare and Medicaid Services. Retrieved from https://www.cms.gov/home/regsguidance.asp

Return On Investment, H. I. P. A. A. (2005). *Blue Cross Blue Shield Association. National Committee on Vital Health Statistics*. Subcommittee on Standards and Security.

Rhee, K., Kwak, J., Kim, S., & Won, D. (2005). Challenge-response based RFID authentication protocol for distributed database environment. In *International Conference on Security in Pervasive Computing* (pp. 70-84). Springer. 10.1007/978-3-540-32004-3_9

Rmayti, M., Khatoun, R., Begriche, Y., Khoukhi, L., & Gaiti, D. (2017). A stochastic approach for packet dropping attacks detection in mobile Ad hoc networks. *Computer Networks*, *121*, 53–64. doi:10.1016/j.comnet.2017.04.027

Rong, B., Chen, H., Qian, Y., Lu, K., Hu, R. Q., & Guizani, S. (2009). A Pyramidal Security Model for Large-Scale Group-Oriented Computing in Mobile Ad Hoc Networks: The Key Management Study. *IEEE Transactions on Vehicular Technology*, *58*(1), 398–408. doi:10.1109/TVT.2008.923666

Rtah, M. (2018). Big Data and IoT-Allied Challenges Associated With Healthcare Applications in Smart and Automated Systems. *International Journal of Strategic Information Technology and Applications*, *9*(2). doi:10.4018/IJSITA.201804010

Rupareliya, J., Vithlani, S., & Gohel, C. (2016). Securing VANET by preventing attacker node using watchdog and Bayesian network theory. *Procedia Computer Science*, *79*, 649–656. doi:10.1016/j.procs.2016.03.082

Russell, S., & Norvig, P. (2010). *Artificial Intelligence: A Modern Approach* (3rd ed.). Upper Saddle River, NJ: Pearson Education.

Sadeghi, K., Banerjee, A., Sohankar, J., & Gupta, S. K. S. (2016). Toward Parametric Security Analysis of Machine Learning based Cyber Forensic Biometric Systems. *IEEE International Conference on Machine Learning and Applications*, 626 – 631. 10.1109/ICMLA.2016.0110

Salah, A. A., & Alpaydin, E. (2004). Incremental mixtures of factor analysers. *Proceedings of the 17th International Conference on Pattern Recognition*. 10.1109/ICPR.2004.1334106

Salem, M. B., & Stolfo, S. J. (2009). *Masquerade attack detection using a search-behavior modeling approach.* Columbia University, Computer Science Department, Technical Report CUCS-027-09.

Santoro, D., Escudero-Andreu, G., Kyriakopoulos, K. G., Aparicio-Navarro, F. J., Parish, D. J., & Vadursi, M. (2017). A hybrid intrusion detection system for virtual jamming attacks on wireless networks. *Measurement*, *109*, 79–87. doi:10.1016/j.measurement.2017.05.034

Saxena, N., Tsudik, G., & Yi, J. H. (2015). Efficient Node Admission and Certificateless Secure Communication in Short-Lived MANETs. *IEEE Transactions on Parallel and Distributed Systems*, *20*(2), 158–170.

Sayan, C. M. (2017). An Intelligent Security Assistant for Cyber Security Operations. *IEEE Conference Publications*, 375 – 376. 10.1109/FAS-W.2017.179

Schuba, C. L., Krsul, I. V., Kuhn, M. G., Spafford, E. H., Sundaram, A., & Zamboni, D. Analysis of a denial of service attack on TCP. In *Security and Privacy, 1997. Proceedings., 1997 IEEE Symposium on* (pp. 208-223). IEEE. 10.1109/SECPRI.1997.601338

Seber, G. A., & Lee, A. J. (2012). *Linear regression analysis* (Vol. 329). John Wiley & Sons.

Sengupta, N., Sen, J., Sil, J., & Saha, M. (2013). Designing of on line intrusion detection system using rough set theory and Q-learning algorithm. *Neurocomputing*, *111*, 161–168. doi:10.1016/j.neucom.2012.12.023

Shah, S. A. R., & Issac, B. (2018). Performance comparison of intrusion detection systems and application of machine learning to Snort system. *Future Generation Computer Systems*, *80*, 157–170. doi:10.1016/j.future.2017.10.016

Shamshirband, S., Amini, A., Anuar, N. B., Kiah, M. L. M., Teh, Y. W., & Furnell, S. (2014). D-FICCA: A density-based fuzzy imperialist competitive clustering algorithm for intrusion detection in wireless sensor networks. *Measurement*, *55*, 212–226. doi:10.1016/j.measurement.2014.04.034

Shamshirband, S., Patel, A., Anuar, N. B., Kiah, M. L. M., & Abraham, A. (2014). Cooperative game theoretic approach using fuzzy Q-learning for detecting and preventing intrusions in wireless sensor networks. *Engineering Applications of Artificial Intelligence*, *32*, 228–241. doi:10.1016/j.engappai.2014.02.001

Shanley, C. W., Jachimowicz, K., & Lebby, M. S. (1994). *Remote retinal scan identifier*. U.S. Patent 5,359,669.

Shareef, A., Zhu, Y., & Musavi, M. (2008). Localization using neural networks in wireless sensor networks. *Proceedings of the 1st International Conference on Mobile Wireless Middleware, Operating Systems, and Applications*, 1–7. 10.4108/ICST.MOBILWARE2008.2901

Sharma & Bhatt. (2018). Privacy Preservation in WSN for Healthcare Application. *Procedia Computer Science, 132*, 1243 – 1252. . doi:10.1016/j.procs.2018.05.040

Shaw, Holway, Alex, Nikolai, Joyce, Hilsenrath, & Speers. (2003). *Method and system for facilitating secure transactions.* U.S. Patent Application 10/032,535.

Shen, G., Zetik, R., Yan, H., Hirsch, O., & Thomä, S. (2011). Time of arrival estimation for range based localization in UWB sensor networks. *Proceedings of IEEE International Conference on Ultra -Wideband, 2,* 1– 4.

Shiu, Y. S., Chang, S. Y., Wu, H. C., Huang, S. C. H., & Chen, H. H. (2011). Physical layer security in wireless networks: A tutorial. *IEEE Wireless Communications, 18*(2), 66–74. doi:10.1109/MWC.2011.5751298

Shi, Y., Xie, W., Xu, G., Shi, R., Chen, E., Mao, Y., & Liu, F. (2003). The smart classroom: Merging technologies for seamless tele-education. *IEEE Pervasive Computing, 2*(2), 47–55. doi:10.1109/MPRV.2003.1203753

Siddiqui, I. F., Qureshi, N. M. F., Shaikh, M. A., Chowdhry, B. S., Abbas, A., Bashir, A. K., & Lee, S. U. J. (n.d.). Stuck-at Fault Analytics of IoT Devices Using Knowledge-based Data Processing Strategy in Smart Grid. *Wireless Personal Communications,* 1-15.

Singh, P. K., Tripathi, B. T., & Singh, N. P. (2011). Node localization in wirless sensor network. *Int J Comput Sci Inform Tech, 26,* 2568–2572.

Soare, C. A. (2012). Internet banking two-factor authentication using smartphones. *Journal of Mobile. Embedded and Distributed Systems, 4*(1), 12–18.

Solove, D. (2013). HIPAA Turns 10: Analyzing the Past, Present, and Future Impact. *Journal of American Health Information Management Association, 84*(4), 22–28.

Song, X., Wu, M., Jermaine, C., & Ranka, S. (2007). Conditional anomaly detection. *IEEE Transactions on Knowledge and Data Engineering, 19*(5), 631–645. doi:10.1109/TKDE.2007.1009

Sperandei, S. (2014). Understanding logistic regression analysis. *Biochemia medica. Biochemia Medica, 24*(1), 12–18. doi:10.11613/BM.2014.003 PMID:24627710

Stamp, M. (2006). *Information Security: Principles and Practice.* Hoboken, NJ: John Wiley & Sons.

Steeves & Snyder. (2007). *Secure online transactions using a CAPTCHA image as a watermark.* U.S. Patent 7,200,576.

Stolfo, S. J., Salem, M. B., & Keromytis, A. D. (2012, May). Fog computing: Mitigating insider data theft attacks in the cloud. In *Security and Privacy Workshops (SPW), 2012 IEEE Symposium on* (pp. 125-128). IEEE.

Surendran & Prakash. (2014). An ACO look-ahead approach to QOS enabled fault-tolerant routing in MANETs. *China Communications*, *12*(8), 93–110.

Sutton, R. S., & Barto, A. G. (2018). *Reinforcement learning: An introduction.* MIT Press.

Swan, M. (2018). Blockchain Economic Networks: Economic Network Theory—Systemic Risk and Blockchain Technology. *Business Transformation through Blockchain*, 3–45. doi:10.1007/978-3-319-98911-2_1

Tao, R., Yang, L., Peng, L., Li, B., & Cemerlic, A. (2009, March). A case study: Using architectural features to improve sophisticated denial-of-service attack detections. In *Computational Intelligence in Cyber Security, 2009. CICS'09. IEEE Symposium on* (pp. 13-18). IEEE.

Tapiador, J. E., & Clark, J. A. (2011). Masquerade mimicry attack detection: A randomised approach. *Computers & Security, 30*(5), 297-310.

Title 45 – Public Welfare, Subtitle A – Department of Health and Human Services, Part 164 – Security and Privacy. (1996). United States. National Archives and Records Administration. Retrieved from http://www.access.gpo.gov/nara/cfr/waisidx_07/45cfr164_07.html

Tran, T. P., Tsai, P., Jan, T., & He, X. (2012). Machine Learning Techniques for Network Intrusion Detection. In Machine Learning: Concepts, Methodologies, Tools and Applications (pp. 498-521). Hershey, PA: IGI Global. doi:10.4018/978-1-60960-818-7.ch310

Tsaih, R., Huang, S., Lian, M., & Huang, Y. (2018). ANN Mechanism for Network Traffic Anomaly Detection in the Concept Drifting Environment. *IEEE Conference on Dependable and Secure Computing (DSC).*

Tu. (1996). Advantages and disadvantages of using artificial neural networks versus logistic regression for predicting medical outcomes. *Journal of Clinical Epidemiology*. Retrieved from https://www.sciencedirect.com/science/article/pii/S0895435696000029

United States Department of Commerce, National Institute of Standards and Technology. (2012). *About NSTIC*. Retrieved from http://www.nist.gov/nstic/about-nstic.html

Van der Aalst, W. M. P. (2016). *Process Mining: Data Science in Action*. Springer. doi:10.1007/978-3-662-49851-4

Verma, A., & Ranga, V. (2018). Statistical analysis of CIDDS-001 dataset for Network Intrusion Detection Systems using Distance-based Machine Learning. *Procedia Computer Science*, *125*, 709–716. doi:10.1016/j.procs.2017.12.091

Vertakova, J., & Plotnikov, V. (2014). Public-private partnerships and the specifics of their implementation in vocational education. *Proceeded Economics and Finance*, *16*, 24–33. doi:10.1016/S2212-5671(14)00770-9

Vimee, W., & Simarpreet, K. (2014). Survey of different localization techniques of wireless sensor network. *IJEEE*, *14*, 759–767.

Walker, S. (2012). Economics and the cyber challenge. *Information Security Technical Report*, *17*(1-2), 9–18. doi:10.1016/j.istr.2011.12.003

Wang, J., & Tao, Q. (2008). Machine Learning: The State of the Art. *IEEE Intelligent Systems*, *23*(6), 49–55. doi:10.1109/MIS.2008.107

Wang, X., Lingjun, G., Shiwen, M., & Santosh, P. (2017). CSI-based fingerprinting for indoor localization: A deep learning approach. *IEEE Trans Veh Tech*, *66*, 763–776.

Wang, Y. (2005). A multinomial logistic regression modeling approach for anomaly intrusion detection. *Computers & Security*, *24*(8), 662–674. doi:10.1016/j.cose.2005.05.003

Wang, Yu, Tang, & Huang. (2009). A Mean Field Game Theoretic Approach for Security Enhancements in Mobile Ad hoc Networks. *IEEE Transactions on Wireless Communications*, *13*(3), 1616–1627.

Watson, H. J. (2014). Tutorial: Big Data Analytics: Concepts, Technologies, and Applications. *Communications of the Association for Information Systems*, *34*, 65. doi:10.17705/1CAIS.03465

Wayman, J., Jain, A., Maltoni, D., & Maio, D. (2005). *An introduction to biometric authentication systems. In Biometric Systems* (pp. 1–20). London: Springer.

Wei, Yingjie, & Mu. (2015). Commercial Bank Credit Risk Evaluation Method based on Decision Tree Algorithm. *IEEE Conference Publications*, 285 -288.

Wei, Z., Tang, H., Yu, F. R., Wang, M., & Mason, P. (2015). Security Enhancements for Mobile Ad Hoc Networks With Trust Management Using Uncertain Reasoning. *IEEE Transactions on Vehicular Technology*, 63(9), 4647–4658.

XGBoost. (2018). In *Wikipedia*. Retrieved from https://en.wikipedia.org/wiki/XGBoost

Xia, Z., & Johnson, M. E. (2010). Access Governance: Flexibility with Escalation and Audit. *Proceedings of the 43rd Hawaii International Conference on System Sciences*.

Xin, Y., Kong, L., Liu, Z., Chen, Y., Li, Y., Zhu, H., ... Wang, C. (2018). Machine Learning and Deep Learning Methods for Cybersecurity. *IEEE Access: Practical Innovations, Open Solutions*, 6, 35365–35381. doi:10.1109/ACCESS.2018.2836950

Xu, W., Huang, L., Fox, A., Patterson, D., & Jordan, M. (2009). Detecting Large-Scale System Problems by Mining Console Logs. *Proceedings of ACM Symposium on Operating Systems Principles (SOSP)* (pp. 117-132). Big Sky, MT: ACM. 10.1145/1629575.1629587

Yang, X. S. (2011). Bat algorithm for multiobjective optimization. *International Journal of Bio-inspired Computation*, 35(5), 267–274. doi:10.1504/IJBIC.2011.042259

Yanping, Z., Daqing, H., & Aimin, J. (2008). Network localization using angle of arrival. *International Conference on Electro Information Technology*, 205–210.

Yavanoglu, O., & Aydos, M. (2017). A review on cyber security datasets for machine learning algorithms. *2017 IEEE International Conference on Big Data (Big Data)*. 10.1109/BigData.2017.8258167

Yi, Y., Wu, J., & Xu, W. (2011). Incremental SVM Based on Reserved Set for Network Intrusion Detection. *Expert Systems with Applications: An International Journal*.

Yick, J., Mukherjee, B., & Ghosal, D. (2008). Wireless sensor network survey. *Computer Networks*, 52, 2292–2330.

Yin, C., Zhang, S., Wang, J., & Kim, J. (2015). An Improved K-Means Using in Anomaly Detection. *First International Conference on Computational Intelligence Theory, Systems and Applications (CCITSA)*. 10.1109/CCITSA.2015.11

Zaitsev, D. A. (2018). Sleptsov Net Computing. In M. Khosrow-Pour (Ed.), Encyclopedia of Information Science and Technology (4th ed.; pp. 7731-7743). Hershey, PA: IGI Global. doi:10.4018/978-1-5225-2255-3.ch672

Zaitsev, D. A. (2019). Sleptsov Net Computing. In M. Khosrow-Pour (Ed.), Advanced Methodologies and Technologies in Network Architecture, Mobile Computing, and Data Analytics (pp. 1660-1674). Hershey, PA: IGI Global. doi:10.4018/978-1-5225-7598-6.ch122

Zarpelao, B. B., Miani, R. S., Kawakani, C. T., & de Alvarenga, S. C. (2017). A survey of intrusion detection in Internet of Things. *Journal of Network and Computer Applications*, *84*, 25–37. doi:10.1016/j.jnca.2017.02.009

About the Contributors

Muhammad Salman Khan is a senior research scientist in cognitive machine intelligence applications in cyber threat hunting and is leading projects in advanced machine learning application in cyber security. Salman has more than a decade of industrial experience in leading research and development projects in cyber security, enterprise networking products and real time data acquisition systems. Salman's Ph.D. research was focussed on improving cognitive performance of machine intelligence algorithms to address the growing challenges of advanced and persistent malware. Salman did his M.S. in Electrical Engineering from Rutgers University USA. Salman has been a recipient of notable national and international awards including Fulbright scholarship, Mitacs Canada fellowship, NSERC PGS (D) scholarship as well as receiving nomination for Canadian Vanier scholarship. He authored and co-authored more than 25 research publications including conference papers, journals and book chapters. Salman is a founder of a technology start-up and has been granted various multi-year research and development funds in technological innovation and development which subsequently led to successful commercialization.

* * *

Rahmoun Abdellatif is a Ph.D. in computer science since 1998. He has been teaching as an assistant professor at the University of Sidi Bel-Abbes up to 2001. He has been enrolled at King Faisal University (Saudi Arabia) as an Associate Professor for eight years. In 2009, He went back to the University of Sidi Bel-Abbes in Algeria, where he got his degree of professor in 2012. Then, he was enrolled at the "High school of Computer science" (ESI-SBA) where he is still working up to now as a full professor. He has been and still being involved in several teaching and research activities in the area of artificial intelligence such as artificial genetics, neural networks, fuzzy logic and fuzzy systems, hybrid intelligent systems, genetic programming, swarm intelligence, and Lately on IoT and embedded systems. Presently, he is the head of research group "Computational Intelligence and Soft Computing (CISCO)" of LaBRi laboratory SBA. His works are published in several research international journals as well as presented in conferences all around the world.

Subrata Acharya is an associate professor in the Computer and Information Sciences department at Towson University. She received her Ph.D. in Computer Science from the University of Pittsburgh, 2008 & M.S. in Computer Engineering from Texas A&M University, College Station, 2004. She has published over 80 peer-reviewed book chapters, research journals and conference publications in the area of computer and information security. Prof. Acharya has obtained significant extramural funding to support her scholarship efforts, including $500K as PI and $250K as co-PI. Of particular note is Prof. Acharya's US patent 7966655 B2, awarded in 2011 with Drs. Wang, Ge and Greenberg for Method and apparatus for optimizing a firewall. Prof. Acharya has also developed new courses in the areas of healthcare informatics, information security and risk management. She has mentored various students who have appeared as co-authors on her papers, and has supervised numerous undergraduate research projects, masters' graduate projects, and doctoral dissertation studies. Her research interests include: Healthcare Information Security and Privacy, Cyber Physical Systems, Medical Analytics, System Security, Trusted Computing, Information Security Management, Secure Health Informatics, Secure Mobile Systems, Ethical & Privacy Issues in Computing Systems.

Rajakumar Arul (S'16 - M'17) is currently associated with Department of Computer Science & Engineering, Amrita School of Engineering, Bengaluru Campus, Amrita Vishwa Vidyapeetham, India as Assistant Professor. He pursued his Bachelor and Master's in Computer Science and Engineering from Anna University, Chennai. Currently, he completed his requirement for Doctorate of Philosophy under the Faculty of Information and Communication in NGNLab, Department of Computer Technology, Anna University - MIT Campus. He is a recipient of the Anna Centenary Research Fellowship for his doctoral studies. His research interests include BlockChain, Security in Broadband Wireless Networks, LTE, Robust resource allocation schemes in Mobile Communication Networks and Cryptography.

Rahul Singh Chowhan pursed Bachelor of Technology from Jodhpur National Univeristy, Jodhpur in 2012, PG-Diploma CDAC from IACSD, Pune in year 2013 and Master of Engineering from M.B.M. Engineering College, Jai Narain Vyas University, Jodhpur in year 2016. He has worked with many MNC's like Webmatrix Tech, Pune and Devlopers Trinity, Jodhpur, He has also been taking guest lectures in Sardar Patel Univeristy of Police, Security and Crime Justice. He is currently working as Senior Research Fellow in Directorate of Extension Education since 2017. He is a life member of International Society for Development & Sustainability since 2017. He has been awarded with academic and scientific awards which includes Young Scientist Award and Excellence in Research award by GECL and Astha Foundation. He has published more than 20 research papers and articles in reputed international journals like Oriental Journal of Computer Science and Technology and has also participated in various International Conferences conducted by IEEE and Springer, it's also available online. His main research work focuses on Mobile Agents, Network Security, Regression testing, Autonomous Load Balancing Mechanisms, Deep Learning, IoT, Parallel Comptuing, Agent-Oriented Distributed Programming and Computational Intelligent Agents based education. He also possess teaching and research experience in his field of interest and has scored CBSE-NET 2018.

Brian Coats is the Assistant Vice President for Technology Operations and Planning at the University of Maryland, Baltimore. He provides both direction and vision as an executive member of the information technology (IT) leadership team for this $1.2 billion institution. In his more than 20 years within higher education, he has proven to be enormously successful in transforming traditional IT units into agile, strategic partners that leverage contemporary tools and approaches to promote collaboration and innovation across the organization. Dr. Coats is a recognized industry expert and active participant in a multitude of national higher education organizations and workgroups related to security and privacy in information technology and his research in these areas has been published and presented both domestically and internationally. He holds a Doctor of Science degree in applied information technology, a Master of Science degree in information technology, an advanced Graduate Certificate in information technology, and a Bachelor of Science degree in aerospace engineering.

Leyla Gamidullaeva graduated from Penza State University, the Faculty of Economics and Management, getting qualifications of an economist. L. Gamidullaeva got her PhD in Economics from Penza State University of Architecture and Construction in 2010 followed by the title of associate professor in 2018. Now she is associate professor at the department of management and economic security of Penza State University. Currently, L. Gamidullaeva is doing her doctoral research in the regional innovation system management at St Petersburg State University. She has authored more than 200 refereed publications and over ten books in innovation management, regional economic growth, networking and collaboration.

Siu Cheung Ho received the Bachelor's Degree in Logistics and Supply Chain Management from University of Northumbria at Newcastle, Master's Degree in Global Supply Chain Management and Master's Degree in Integrated Engineering from The Hong Kong Polytechnic University. Before he started his Engineering Doctorate Degree studies, he was a Manager, Program & Projects Operations at Hong Kong R&D Centre for Logistics and Supply Chain Management Enabling Technologies (LSCM). His current research interests include innovation and technology engineering management, system engineering and project management, vision-based analytics, robotic, big data and pervasive computing. He is a committee member in Education and Training Committee & Logistics Policy Committee and Chartered Member of The Chartered Institute of Logistics and Transport in Hong Kong (CILTHK). He is Technical Advisor in The Employees Retraining Board (ERB).

Ali Kashif Bashir is a Senior Lecturer at School of Computing, Mathematics, and Digital Technology, Manchester Metropolitan University, United Kingdom. He is a senior member of IEEE and Distinguished Speaker of ACM. His past assignments include: Associate Professor of Information and Communication Technologies, Faculty of Science and Technology, University of the Faroe Islands, Denmark; Osaka University, Japan; Nara National College of Technology, Japan; the National Fusion Research Institute, South Korea; Southern Power Company Ltd., South Korea, and the Seoul Metropolitan Government, South Korea. He received his Ph.D. in computer science and engineering from Korea University, South Korea. MS from Ajou University, South Korea and BS from University of Management and Technology, Pakistan. He is author of over 80 peer-reviewed articles. He is supervising/co-supervising several graduate (MS and PhD) students. His research interests include internet of things, wireless networks, distributed systems, network/cyber security, cloud/network function virtualization, etc. He is serving as the Editor-in-chief of the IEEE FUTURE DIRECTIONS NEWSLETTER. He is editor of several journals and also has served/serving as guest editor on several special issues in journals of IEEE, Elsevier, and Springer. He has served as chair (program, publicity, and track) chair on several conferences and workshops. He has delivered several invited and keynote talks, and reviewed the technology leading articles for journals like IEEE TRANSACTIONS ON INDUSTRIAL INFORMATICS, the IEEE Communication Magazine, the IEEE COMMUNICATION LETTERS, IEEE Internet of Things, and the IEICE Journals, and conferences, such as the IEEE Infocom, the IEEE ICC, the IEEE Globecom, and the IEEE Cloud of Things.

Pascal Lorenz received his M.Sc. (1990) and Ph.D. (1994) from the University of Nancy, France. Between 1990 and 1995 he was a research engineer at WorldFIP Europe and at Alcatel-Alsthom. He is a professor at the University of Haute-Alsace, France, since 1995. His research interests include QoS, wireless networks and high-speed networks. He is the author/co-author of 3 books, 3 patents and 200 international publications in refereed journals and conferences. He was Technical Editor of the IEEE Communications Magazine Editorial Board (2000-2006), Chair of Vertical Issues in Communication Systems Technical Committee Cluster (2008-2009), Chair of the Communications Systems Integration and Modeling Technical Committee (2003-2009) and Chair of the Communications Software Technical Committee (2008-2010). He has served as Co-Program Chair of IEEE WCNC'2012, ICC'2004 and ICC'2017, tutorial chair of VTC'2013 Spring and WCNC'2010, track chair of PIMRC'2012, symposium Co-Chair at Globecom 2007-2011, ICC 2008-2010,

Miloud Mihoubi is younger researcher in computer sciences, He received his engineering degree in Computer Science in 2009 from the Computer Science department of University of sciences and technology Oran Mohamed Boudiaf (USTO- MB) - Algeria, in 2013, he received the Magister diploma in Artificial Intelligence and pattern recognition from the USTO- MB – Algeria . Miloud is a senior member in Evolutionary Engineering and Distributed Information System (EEDIS) in Computer Science Department Exact Science Faculty Djillali Liabes University at Sidi Bel Abbes (SBA). He is also member researcher at Computational Intelligence and Soft Computing (CISCO) of LaBRi laboratory SBA university. He has been and still being involved in several teaching and research activities in several areas such as Wireless Sensor Network, Optimization Problem, Cybersecurity, Deep Learning and internet of thing.

Sushruta Mishra is currently working as Asst. Professor at KIIT University. His research interests includes machine learning and data mining.

Vardan Mkrttchian received his Doctorate of Sciences (Engineering) in Control Systems from Lomonosov Moscow State University (former USSR). Dr. Vardan Mkrttchian taught for undergraduate and graduate students courses of control system, information sciences and technology, at the Astrakhan State University (Russian Federation), where he was is the Professor of the Information Systems (www.aspu. ru) six years. Now he is full professor in CAD department of Penza State University (www.pnzgu.ru). He is currently chief executive of HHH University, Australia and team leader of the international academics (www.hhhuniversity.com). He also serves as executive director of the HHH Technology Incorporation. Professor Vardan Mkrttchian has authored over 400 refereed publications. He is the author of over twenty books published of IGI Global, included ten books indexed of SCOPUS in IT, Control System, Digital Economy, Education Technology. He is also has authored more than 200 articles published in various conference proceedings and journals. From October 20, 2016 Prof. V. Mkrttchian is Director of Center Intellectual Control System of International University in Moscow City (www.interun.ru).

Melody Moh obtained her MS and Ph.D., both in computer science, from Univ. of California - Davis. She joined San Jose State University in 1993, and has been a Professor since Aug 2003. Her research interests include cloud computing, software defined networks, mobile, wireless networking, and security/privacy for cloud and network systems. She has received over 500K dollars of research grants from both NSF and industry, has published over 150 refereed papers in international journals, conferences and as book chapters, and has consulted for various companies.

Svetlana Panasenko received higher economic education in 1992 at the Stavropol Polytechnic Institute with a degree in Economics and management, from 1992 to 1996 she worked as a chief economist in business organizations, in 2007 she received a degree of doctor of economic Sciences at the North Caucasus state technical University, where from 2007 to 2010 she worked as a Professor at the faculty of Economics and Finance. In 2015, she received a master's degree in practical psychology from Moscow state University of psychology and education. Since 2010 she has been accepted to the position of Professor at the Plekhanov Russian state University in Moscow and currently holds the position of head of the Department of trade policy at this University. She writes and widely represents the issues of e-Commerce, Neurotechnology, marketing and management in the digital economy. She is the author of works on the analysis of export-import relations between Russia and other countries, innovative approaches to the development of various sectors of the economy (Espacios, 2018).

Mamata Rath, M.Tech, Ph.D (Comp.Sc), has twelve years of experience in teaching as well as in research and her research interests include Mobile Adhoc Networks, Internet of Things, Ubiquitous Computing, VANET and Computer Security.

Rajalakshmi Shenbaga Moorthy received her B.Tech under the stream of Information Technology from Mookambigai College of Engineering in 2010. She completed her Master of Engineering (M.E) in the stream of Computer Science and Engineering from Madras Institute of Technology, Anna University in 2013. She is the Gold Medalist in her B.Tech and M.E programme. She is currently working as an Assistant Professor in St. Joseph's Institute of Technology, Chennai. Her research area includes Analysis of Algorithms, Machine learning, Resource Management Techniques. She has published various International conference papers and Scopus indexed journals

Rohit Tanwar is currently working as Assistant Professor (Senior Scale) in University of Petroleum, Energy and Studies, Dehradun, INDIA. He is pursuing his Ph.D in computer engineering from University Institute of Engineering and Technology, Kurukshetra University, Kurukshetra (INDIA). Mr. Tanwar has completed his M.Tech in computer engineering from YMCA University of Science & Technology, Faridabad (INDIA). He is having more than 10 years of experience in teaching. He is currently doing research in the field of Network Security. He has around fifteen publications to his credit in different reputed journals and conferences. He has conducted special sessions in various national and international conferences.

Yulia Vertakova, D.Sc. (Economy), is Head of Department of regional economics and management of Southwest State University, Russian Federation. She is the leading scientist of digital economy, industry development 4.0, indicative governance of sustainable development of regional economy, cluster initiatives in the region, structural transformation industrial complex regulation in terms of digitalization of economy, green economy, scaling business models, business planning, innovative management, strategic management, socio-economic systems, indicative planning, regional economics, socio-economic forecasting, proactive management, reengineering of business processes.

Yuen Kwan Yau received the Bachelor's Degree in Investment Science from The Hong Kong Polytechnic University. Currently, Ms Yau is a master student in Information Technology.

Steven Yen is a graduate student at the Dept. of Computer Science, San Jose State University. His research interests include applying machine learning and deep learning techniques to cybersecurity and cloud computing.

Chi Kwan Yip received the Bachelor's Degree in Electronic and Information Engineering from The Hong Kong Polytechnic University. He devoted himself to IT Industry over 15 years in different regions, such as Manufacturing, Medical and Retail. And he became one of the key managerial people in recent years.

Index

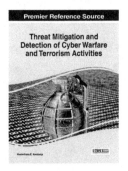

Ensure Quality Research is Introduced to the Academic Community

Become an IGI Global Reviewer for Authored Book Projects

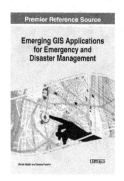
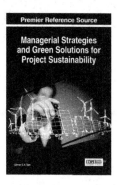

The overall success of an authored book project is dependent on quality and timely reviews.

In this competitive age of scholarly publishing, constructive and timely feedback significantly expedites the turnaround time of manuscripts from submission to acceptance, allowing the publication and discovery of forward-thinking research at a much more expeditious rate. Several IGI Global authored book projects are currently seeking highly qualified experts in the field to fill vacancies on their respective editorial review boards:

Applications may be sent to:
development@igi-global.com

Applicants must have a doctorate (or an equivalent degree) as well as publishing and reviewing experience. Reviewers are asked to write reviews in a timely, collegial, and constructive manner. All reviewers will begin their role on an ad-hoc basis for a period of one year, and upon successful completion of this term can be considered for full editorial review board status, with the potential for a subsequent promotion to Associate Editor.

If you have a colleague that may be interested in this opportunity,
we encourage you to share this information with them.

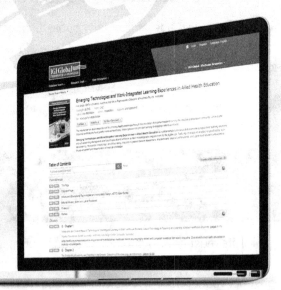

Printed in the United States
By Bookmasters